Errata

Page	Line/Eq./Fig.	Replace	By
29	Line 4	Minkowski	Minkowski space-time
133	Line 10 from the bottom	See diagram below	See Fig. 6.2
139	Eq.(6.6)	$x = (z + z^*)$	$x = (z + z^*)/2$
139	Eq.(6.7)	$i[v_x + u_y]$	$i\frac{1}{2}\left[v_x + u_y\right]$
165	Fig.7.3*a* caption	Gursat	Goursat
166	Fig.7.3*b* caption	Gursat	Goursat
203	Line 5 from the bottom	...is analytic on and within C	...is analytic on and within C (except for pole at $z = -1$)
228	Caption of Fig. 8.	e^{t}	e^{-t}
228	Caption of Fig. 8.	$e^{t} t^{w}$	$e^{-t} t^{w}$
229	Line 2	dw	dx
236	Line 12	vicimity	vicinity
237	Lines 16, 18 and 20	$(1+i)$	$2\,(1+i)$

MATHEMATICAL PHYSICS

THE BASICS

S D Joglekar (Late)

Professor, Department of Physics
Indian Institute of Technology Kanpur
India

Universities Press

MATHEMATICAL PHYSICS: THE BASICS

UNIVERSITIES PRESS (INDIA) PRIVATE LIMITED

Registered Office
3-6-747/1/A & 3-6-754/1 Himayatnagar, Hyderabad 500 029, Telangana, India
info@universitiespress.com; www.universitiespress.com

Distributed by
Orient Blackswan Private Limited

Registered Office
3-6-752 Himayatnagar, Hyderabad 500 029, Telangana, India

Other Offices
Bengaluru, Chennai, Guwahati, Hyderabad, Kolkata,
Mumbai, New Delhi, Noida, Patna

First published 2005
Reprinted 2011, 2017, 2024, 2025

502206

Typeset by
Krishtel eMaging Solutions Pvt. Ltd., Chennai 600 017

ISBN 978 81 7371 422 1

Printed in India at
Avantika Printers Private Limited, New Delhi 110 020

Published by
Universities Press (India) Private Limited
3-6-747/1/A & 3-6-754/1 Himayatnagar, Hyderabad 500 029, Telangana, India
info@universitiespress.com; www.universitiespress.com

Contents

Preface

Understanding advanced physics requires a thorough grasp of the mathematical basis of physics. The present-day theoretical physics involves a large number of basic mathematical techniques. This two-volume set on mathematical physics aims to cover these areas that are essential at the Master's level in physics and chemistry. Written keeping in mind the needs of post-graduate students of physics in Indian universities, these books will also be useful to post-graduate students of chemistry and to engineering students. This is also the minimum underlying preparation needed in mathematics for research by a student of either physics or chemistry.

The contents of the volumes have been determined by my experience of teaching such courses at IIT Kanpur, and the level adjusted appropriately for a two-year MSc program. In the first volume, *Mathematical Physics: The Basics*, the focus is on developing the basic topics such as vectors and tensors, Cartesian tensors, Lorentz tensors, curvilinear coordinates, linear vector spaces, linear operators on linear vector spaces, matrices, complex variables and applications at an advanced level. A kKnowledge of these is essential to grasp the advanced and more specialised topics that are dealt with in the second volume *Mathematical Physics: Advanced Topics*.

It is assumed that the student is familiar with the basic operations in vector algebra and vector calculus. The topic of vector algebra and analysis is best treated in a condensed summation convention notation using Levi–Civita and Kronecker delta symbols. This, moreover, paves the way for their logical extension to tensor analysis. These are covered in Chapter 1. Chapter 2 deals briefly with Lorentz transformations and Lorentz tensors, essential for an understanding of the special theory of relativity and classical electrodynamics (as also the branches of physics that use relativity such as high-energy physics and general relativity). Chapter 3 deals with orthogonal curvilinear coordinates and their use in vector calculus. Chapters 4 and 5 together cover at length the linear vector spaces and the linear operators. They also introduce vector subspaces, eigenvalues and eigenvectors, similarity transformations, matrix representations and properties of a class of matrices relevant to physics and their diagonalization. Chapters 6, 7 and 8 cover the topic of the complex variables and their various applications. Chapter 6 introduces the notion of an analytic function and several special analytic functions. Chapter 7 deals with contour integrals, the Cauchy–Goursat theorem, the Taylor and the Laurent series, the residue theorems and its applications. Chapter 8 deals with the analytic continuation, conformal mapping and its applications to electrostatics. the method of steepest descent and Riemann surfaces.

In this book, an attempt has been made to make the connection of the mathematical topics with physics clear. Concepts are introduced and results established rigorously and also elaborated intuitively. A large number of solved examples and a large number of exercises on every topic are included. It is hoped that the book fulfils the need of the MSc/Engineering students and their instructors. The topics, as stated earlier, form the backbone of all the mathematical techniques needed by physicist, chemists and engineers, and as such will be valuable to students of all these disciplines.

The second volume contains several other topics relevant to physics such as partial differential equations, ordinary differential equations, special functions, orthogonal polynomials, generalized functions, Fourier series, and elements of group theory. The two volumes together provide the material necessary for a two-semester course on mathematical methods in physics. This volume can be used for the first-semester course.

This book was supported financially by Quality Improvement Program at IIT Kanpur. I would also like to express my gratitude to Professor Vijay A Singh of IIT Kanpur for his critical reading of several chapters of the book, his suggestions and his appreciative comments. I would also like to thank Ghanashyam Hoshing for the typing of the manuscript.

IIT Kanpur
March 2004

S D Joglekar

1 Coordinate Transformations, Vectors and Cartesian Tensors

We are familiar with the fact that the laws of physics are generally expressed in terms of quantities called scalars, vectors and tensors. This chapter gives the definitions of these objects in terms of their properties under coordinate transformations. To this end, we shall first discuss coordinate transformations in two- and three-dimensions, and then discuss vectors and cartesian tensors. Before we do this, it is however natural to ask the question as to why is it that such objects naturally arise in physical laws.

To understand the significance of the concepts of scalars, vectors and tensors in physics and the reasons for their importance in physics, it is first necessary to draw attention to the fundamental assumptions underlying the formulations of *nonrelativistic* classical physical theories (Newtonian mechanics, electrostatics, etc.). These are:

1. A 3-dimensional Euclidean space and an independent time, definitions of which are prior to and independent of a particular physical phenomenon one is studying.
2. Existence of a physical phenomenon that is independent of its description in terms of a coordinate frame chosen to describe space and time. (The coordinate frames can, of course, be chosen variously, and the corresponding descriptions of the same phenomenon may vary accordingly (see Problem 1.23).
3. A fundamental fact found from experience (and hence raised to the level of a postulate) that the fundamental physical laws can be described in a mathematical form which is independent of the coordinate frame chosen.

As a result, it stands to reason that the quantities involved in the formulation of natural laws are represented by mathematical entities that have "simple" transformation properties under transformation of coordinates. Such quantities are classified, according to their specific properties, as scalars, vectors and tensors. For example, the inertial property of a body in relation to a purely translational motion is represented by "mass" which is invariant under coordinate rotations, i.e., a scalar. The electric force on an electric charge is represented by a vector. The inertial properties of a rigid body under arbitrary rotations are represented by a tensor of rank two (the inertia tensor), and so on.

1.1 Coordinate Transformations

Consider the cartesian system of coordinates 'S' in two-dimensions with origin at O. Consider another cartesian system S$'$ having the same origin O but with axes rotated through an angle θ

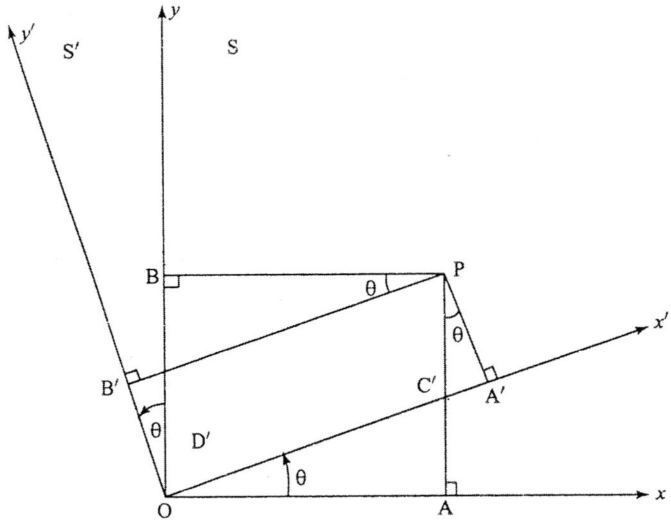

Fig. 1.1 *Rotation of Cartesian coordinates in two dimensions*

as shown in Fig. 1.1. Consider any point P in the $(x-y)$ plane. The same physical point P will be described by coordinates (x, y) in S and by (x', y') in S$'$. We want to obtain the relation between these two sets of coordinates of the same point P in the two coordinate system.

$$\mathbf{OP} = \mathbf{r} = x\hat{\boldsymbol{\varepsilon}}_1 + y\hat{\boldsymbol{\varepsilon}}_2 \tag{1.1a}$$

$$= x'\hat{\boldsymbol{\varepsilon}}'_1 + y'\hat{\boldsymbol{\varepsilon}}_2 \tag{1.1b}$$

$$x' = \mathbf{r}\cdot\hat{\boldsymbol{\varepsilon}}'_1 = x\hat{\boldsymbol{\varepsilon}}_1\cdot\hat{\boldsymbol{\varepsilon}}'_1 + y\hat{\boldsymbol{\varepsilon}}_2\cdot\hat{\boldsymbol{\varepsilon}}'_1 \tag{1.2a}$$

$$\therefore \quad x' = x\cos\theta + y\sin\theta \tag{1.3a}$$

$$y' = \mathbf{r}\cdot\hat{\boldsymbol{\varepsilon}}'_2 = x\hat{\boldsymbol{\varepsilon}}_1\cdot\hat{\boldsymbol{\varepsilon}}'_2 + y\hat{\boldsymbol{\varepsilon}}_2\cdot\hat{\boldsymbol{\varepsilon}}'_2 \tag{1.2b}$$

$$\therefore \quad y' = -x\sin\theta + y\cos\theta \tag{1.3b}$$

From (1.3a) and (1.3b), we obtain, in matrix notation

$$X' \equiv \begin{bmatrix} x' \\ y' \end{bmatrix} = \begin{bmatrix} \cos\theta & \sin\theta \\ -\sin\theta & \cos\theta \end{bmatrix} \begin{bmatrix} x \\ y \end{bmatrix} \equiv A(\theta)X, \tag{1.4}$$

where the square matrix

$$A(\theta) \equiv \begin{bmatrix} \cos\theta & \sin\theta \\ -\sin\theta & \cos\theta \end{bmatrix} \tag{1.5}$$

is the matrix for coordinate rotations in two-dimensions.

Relation (1.3a) and (1.3b) can be inverted to obtain x, y in terms of x', y'.

These are, in matrix notation,

$$X \equiv \begin{bmatrix} x \\ y \end{bmatrix} = \begin{bmatrix} \cos\theta & -\sin\theta \\ \sin\theta & \cos\theta \end{bmatrix} \begin{bmatrix} x' \\ y' \end{bmatrix} = A^{\mathrm{T}}(\theta)X'. \tag{1.6}$$

Thus, the inverse transformation involves transpose of $A(\theta)$. It is easy to verify that

$$A(\theta)A^{\mathrm{T}}(\theta) = A^{\mathrm{T}}(\theta)A(\theta) = \mathbb{1} \tag{1.7}$$

where $\mathbb{1}$ is the 2×2 identity matrix, indicating that $A(\theta)$ is an orthogonal matrix.

Observations: We have so far considered only a rotation of cartesian coordinates in two-dimensions. We will now observe a number of features of the transformations (1.6) and (1.7), and then note which of these remain valid in higher dimensions.

(*a*) $A(\theta)$ is independent of x, y, i.e., the rotation matrix $A(\theta)$ is independent of the point P.
(*b*) $A(\theta)$ depends only on one angle θ.
(*c*) $A(\theta)$ is nonsingular. This must be so if the inverse transformation matrix $A^{-1}(\theta)$ is to exist.
(*d*) $A(\theta)$ is, in addition, orthogonal.
(*e*) $\det A(\theta) = +1$ (see Problem 1.2).
(*f*) Two successive coordinate transformations can be applied in any order giving the same result:

$$A(\theta')A(\theta) = A(\theta)A(\theta'),$$

i.e., rotation matrices in two-dimensions are commutative.
(*g*) Every 2×2 orthogonal matrix with $\det A = +1$ corresponds to some rotation through an angle θ (see Problem 1.4).

We now comment on whether these features are preserved in higher dimensions, say n, the motive for which follows. We certainly need to use coordinate transformations in the 3-dimensional space we live in. Lorentz transformations in Special Theory of Relativity can be looked upon as rotations in 4-dimensional space–time.

Property (*a*) is valid for a coordinate transformation in any-dimensional Euclidean space provided the origin is fixed and the coordinate system is cartesian. This property arises from the homogeneity of space.

Similarly, properties (*c*) and (*d*) remain valid for the n-dimensional Euclidean space.

Regarding property (*b*), the rotation matrix A will generally depend on as many "independent rotations" as there are in n-dimensions. This number turns out to be $n(n-1)/2$.

Property (*e*) is also valid in any dimensions (for proper rotations).

The property (f) is specific only to two-dimensions and arises from the fact that there is only one axis of rotation there. In ($n \geq 3$)-dimensional Euclidean space, there are more than one independent rotations (such as in 3-dimensions, there are three independent rotations about each of the x, y and z axes), and such (finite) rotations do not (generally) commute (see Problems 1.8 and 1.9).

Property (g) also holds in three-dimensions. Any orthogonal matrix with det $A = +1$ is in fact a rotation through some angle about some axis.

1.1.1 Coordinate transformations in 3-dimensions

We live in a 3-dimensional space. So, the discussion of rotations of coordinate frames in two-dimensions considered in the previous subsection is relevant only as a special case of the more general rotations of coordinate frames in 3-dimensions, and was given to introduce the general properties of coordinate transformations under rotations. In this subsection we shall first discuss in a general way an arbitrary coordinate rotation in 3-dimensions. Then, we shall see how they can be described in terms of the Euler angles.

We proceed much the same way as in Section 1.1. Let S and S′ be two Cartesian coordinate systems in 3-dimensional space with a common origin O. Let P be an arbitrary point in space. In terms of the orthogonal triads $\hat{\boldsymbol{\varepsilon}}_1$, $\hat{\boldsymbol{\varepsilon}}_2$, $\hat{\boldsymbol{\varepsilon}}_3$ and $\hat{\boldsymbol{\varepsilon}}'_1$, $\hat{\boldsymbol{\varepsilon}}'_2$, $\hat{\boldsymbol{\varepsilon}}'_3$ of the two coordinate systems, the position vector of P can be resolved as

$$\mathbf{OP} = \mathbf{r} = x\hat{\boldsymbol{\varepsilon}}_1 + y\hat{\boldsymbol{\varepsilon}}_2 + z\hat{\boldsymbol{\varepsilon}}_3 \tag{1.8a}$$

$$= x'\hat{\boldsymbol{\varepsilon}}'_1 + y'\hat{\boldsymbol{\varepsilon}}'_2 + z'\hat{\boldsymbol{\varepsilon}}'_3. \tag{1.8b}$$

We obtain the relations between (x', y', z') and (x, y, z), the coordinates of P in S and S′ respectively, as before.

$$x' = \mathbf{r}\cdot\hat{\boldsymbol{\varepsilon}}'_1 = \left[x\hat{\boldsymbol{\varepsilon}}_1 + y\hat{\boldsymbol{\varepsilon}}_2 + z\hat{\boldsymbol{\varepsilon}}_3\right]\cdot\hat{\boldsymbol{\varepsilon}}'_1$$

$$= \left[\hat{\boldsymbol{\varepsilon}}_1\cdot\hat{\boldsymbol{\varepsilon}}'_1\right]x + \left[\hat{\boldsymbol{\varepsilon}}_2\cdot\hat{\boldsymbol{\varepsilon}}'_1\right]y + \left[\hat{\boldsymbol{\varepsilon}}_3\cdot\hat{\boldsymbol{\varepsilon}}'_1\right]z \tag{1.9a}$$

In a similar fashion,

$$y' = \left[\hat{\boldsymbol{\varepsilon}}_1\cdot\hat{\boldsymbol{\varepsilon}}'_2\right]x + \left[\hat{\boldsymbol{\varepsilon}}_1\cdot\hat{\boldsymbol{\varepsilon}}'_2\right]y + \left[\hat{\boldsymbol{\varepsilon}}_3\cdot\hat{\boldsymbol{\varepsilon}}'_2\right]z, \tag{1.9b}$$

$$z' = \left[\hat{\boldsymbol{\varepsilon}}_1\cdot\hat{\boldsymbol{\varepsilon}}'_3\right]x + \left[\hat{\boldsymbol{\varepsilon}}_2\cdot\hat{\boldsymbol{\varepsilon}}'_3\right]y + \left[\hat{\boldsymbol{\varepsilon}}_3\cdot\hat{\boldsymbol{\varepsilon}}'_3\right]z. \tag{1.9c}$$

We can write (1.9) in a matrix form

$$\begin{bmatrix} x' \\ y' \\ z' \end{bmatrix} = \begin{bmatrix} \hat{\boldsymbol{\varepsilon}}_1\cdot\hat{\boldsymbol{\varepsilon}}'_1 & \hat{\boldsymbol{\varepsilon}}_2\cdot\hat{\boldsymbol{\varepsilon}}'_1 & \hat{\boldsymbol{\varepsilon}}_3\cdot\hat{\boldsymbol{\varepsilon}}'_1 \\ \hat{\boldsymbol{\varepsilon}}_1\cdot\hat{\boldsymbol{\varepsilon}}'_2 & \hat{\boldsymbol{\varepsilon}}_2\cdot\hat{\boldsymbol{\varepsilon}}'_2 & \hat{\boldsymbol{\varepsilon}}_2\cdot\hat{\boldsymbol{\varepsilon}}'_3 \\ \hat{\boldsymbol{\varepsilon}}_1\cdot\hat{\boldsymbol{\varepsilon}}'_1 & \hat{\boldsymbol{\varepsilon}}_2\cdot\hat{\boldsymbol{\varepsilon}}'_2 & \hat{\boldsymbol{\varepsilon}}_3\cdot\hat{\boldsymbol{\varepsilon}}'_3 \end{bmatrix} \begin{bmatrix} x \\ y \\ z \end{bmatrix}. \tag{1.10}$$

These read in an obvious notations (see analogous equation (1.4))

$$X' = AX \tag{1.11}$$

with

$$A_{ij} = \hat{\boldsymbol{\varepsilon}}_i \cdot \hat{\boldsymbol{\varepsilon}}'_j. \tag{1.12}$$

We could have started the other way and obtained X in terms of X'. Then we would have

$$X = BX' \tag{1.13}$$

with

$$B_{ij} = \hat{\boldsymbol{\varepsilon}}'_i \cdot \hat{\boldsymbol{\varepsilon}}_j. \tag{1.14}$$

The matrix A is orthogonal for

$$(A^{\mathrm{T}}A)_{kj} = \sum_i A^{\mathrm{T}}_{ki}\, A_{ij} = \sum_i A_{ik}\, A_{ij}$$

$$= \sum_i \hat{\boldsymbol{\varepsilon}}_i \cdot \hat{\boldsymbol{\varepsilon}}'_k \; \hat{\boldsymbol{\varepsilon}}_i \cdot \hat{\boldsymbol{\varepsilon}}'_j$$

$$= \left[\sum_i \hat{\boldsymbol{\varepsilon}}'_k \cdot \hat{\boldsymbol{\varepsilon}}_i \quad \hat{\boldsymbol{\varepsilon}}_i \right] \cdot \hat{\boldsymbol{\varepsilon}}'_j \tag{1.15}$$

$$= \hat{\boldsymbol{\varepsilon}}'_k \cdot \hat{\boldsymbol{\varepsilon}}'_j = \delta_{kj} = (\mathbb{1})_{kj} \tag{1.16}$$

where, in going from (1.15) to (1.16) we have used that for any vector $\mathbf{C}$,

$$\mathbf{C} = \sum_i \mathbf{C} \cdot \hat{\boldsymbol{\varepsilon}}_i \; \hat{\boldsymbol{\varepsilon}}_i.$$

Thus $A^{\mathrm{T}}A = \mathbb{1}$; proving that A is an orthogonal matrix. (One may also verify by the same procedure that $A^{\mathrm{T}}A = \mathbb{1}$. Of course, it is implied by the relation that $A^{\mathrm{T}}A = \mathbb{1}$ for any finite–dimensional A.) Of course, then A^{-1} exists and from Eq. (1.13), $B = A^{-1} = A^{\mathrm{T}}$.

We shall now proceed to study the properties of the rotation matrix A in analogy with those of the two-dimensional rotational matrices considered in Section 1.1. Some of the statements will not be proved here but will find natural proofs elsewhere in the book (as indicated in the respective places).

(*a*) The rotation matrix A is independent of x, y, z, i.e., of the point P.

(*b*) A, however, now depends on *three* independent parameters. This is seen from the fact that the nine elements of A are subject to the matrix condition

$$A^{\mathrm{T}}A = \mathbb{1}$$

which, as shown below, corresponds to six independent conditions. The above matrix condition amounts to

$$(i)\quad (A^{\mathrm{T}}A)_{ii} = \sum_j a_{ji}\, a_{ji} = \sum_j a_{ji}^2 = 1 \qquad i = 1, 2, 3$$

$$(ii)\quad (A^{\mathrm{T}}A)_{ij} = \sum_{k=1}^{3} a_{ki}\, a_{kj} = (A^{\mathrm{T}}A)_{ji} = 0 \qquad (i, j) = (1\ 2), (1\ 3), (2\ 3).$$

There are three independent conditions in (i) and (ii) each; thus six conditions in all. The existence of three independent parameters is alternately understandable since now there are three independent rotations, one in each of the x–y, y–z and z–x planes, or here, one about each axis.

(c) In view of the fact that $A^{\mathrm{T}}A = \mathbb{1}$, A^{-1} exists and is A^{T}.

(d) A is orthogonal.

(e) $A^{\mathrm{T}}A = \mathbb{1} \Rightarrow \det A^{\mathrm{T}} \det A = (\det A)^2 = 1.$
$\det A = \pm 1.$

For a pure rotation $\det A$ must be $+1$. This can be seen as follows: A pure rotation is characterized by an axis $\hat{\mathbf{n}}$ of rotation and an angle of rotation θ. The rotation through θ can be understood as an effect of N rotations of $\frac{\theta}{N}$ each about the same axis. $A_{\hat{\mathbf{n}}}(\theta)$ is constructed as

$$X' = A_{\hat{\mathbf{n}}}(\theta)X = \left[A_{\hat{\mathbf{n}}}\left(\frac{\theta}{N}\right)\right]^N X$$

so that

$$\det A_{\hat{\mathbf{n}}}(\theta) = \left[\det A_{\hat{\mathbf{n}}}\left(\frac{\theta}{N}\right)\right]^N$$

Now, $A_{\hat{\mathbf{n}}}\left(\frac{\theta}{N}\right)$ is a pure rotation itself and must have determinant $+1$ or -1. But as $N \to \infty$, $\frac{\theta}{N}$ is infinitesimal and $\det A_{\hat{\mathbf{n}}}\left(\frac{\theta}{N}\right)$ can differ from $\det A_{\hat{\mathbf{n}}}(0)$ only by an infinitesimal amount. (Continuity of $\det A_{\hat{\mathbf{n}}}\left(\frac{\theta}{N}\right)$ in θ has been utilized.) But $A_{\hat{\mathbf{n}}}(0) = \mathbb{1}$, as it corresponds to no rotation at all. So $\det A_{\hat{\mathbf{n}}}(0) = +1$.

Hence

$$\det A_{\hat{\mathbf{n}}}\left(\frac{\theta}{N}\right) = 1 + \text{an infinitesimal quantity.}$$

But the left-hand side can only take discrete values ± 1. Hence it must be $+1$. (To put it qualitatively, determinant for this rotation cannot suddenly jump from $+1$ to -1 assuming continuity of $A(\theta/N)$ on θ.)

Thus,

$$\det A_{\hat{\mathbf{n}}}(\theta) = (+1)^N \qquad \text{as } N \to \infty$$
$$= 1.$$

Thus a pure rotation is always characterized by a matrix with determinant +1. In fact every orthogonal real A with $\det A = +1$ corresponds to a pure rotation (see Problem 1.4 and comments after Theorem 5.13). Such a matrix is said to represent a *proper* rotation.

We note further that the condition $A^{\mathrm{T}}A = \mathbb{1}$ on the coordinate transformation matrix A does allow A's with $\det A = -1$. These then cannot correspond to what we ordinarily understand by a rotation. These are called *improper rotations*. A simple example of A with $\det A = -1$ is $A = \mathrm{diag}(-1, -1, -1)$. This A reverses the sense of all the three coördinate axes and therefore takes one from a right-handed coordinate system to a left-handed coordinate system. We will denote this matrix by P (as it represents what is known as the parity transformation). In fact, if $\det A = -1$, then (AP) represents a pure rotation. To see this, note that $\det (AP) = \det A \det P = +1$; and also that $(AP)^{\mathrm{T}}(AP) = P^{\mathrm{T}}A^{\mathrm{T}}AP = P^{\mathrm{T}}P = \mathbb{1}$. Thus every improper rotation A can be represented as

$$A = (AP)P = \text{A pure rotation } (AP) \times \text{Parity transformation}$$

(f) In three-dimensions, generally the order in which two successive rotations about different axes are applied matters. This is most easily demonstrated with the help of a simple example. Suppose we apply two rotations through $\pi/2$ each in two different orders about x and y axes (say). It is then easily verified by a visual construction that

$$R_x(\pi/2)\ R_y(\pi/2) \neq R_y(\pi/2)\ R_x(\pi/2).$$

Thus two rotations about different axes generally do not commute unless both are infinitesimal ones. (For exceptions see Problem 1.8.)

1.2 Euler Angles

As seen earlier, a general rotation depends on three independent variable parameters. There may be many different ways of choosing these three independent parameters (see Problem 1.6); but one particular choice (and its minor variations) defined in terms of the so-called "Euler angles" is used most commonly. In this approach, we use the fact that the coordinate rotation about a coordinate axis (an "elementary rotation") takes a particularly simple form and depends on only one angle. We then express a general rotation (about some axis through some angle) as a succession of three "elementary rotations" about certain coordinate axes (to be specified), each depending on one angle. Thus an arbitrary 3-D rotation is expressed as a succession of three particular 2-D (elementary) rotations. The expression so obtained is very useful in quantum mechanics and is also used in describing the dynamics of rotating rigid bodies in classical mechanics.

Suppose we want to represent an arbitrary rotation specified by an axis $\hat{\mathbf{n}}$ and an angle θ as shown in the Fig. 1.2. We shall show how this rotation can be represented as a series of three rotations each about a coordinate axis. This is done as follows. To represent the rotation about ON as a rotation about an axis, we must bring one of the axes (say the z-axis) along ON by a rotation of the coordinate system. This is accomplished (in particular) by rotating the coordinate system about an axis through O normal to the plane defined by ON and the z-axis through an angle β as shown. This axis is of course perpendicular to the z-axis and through O, and hence lies in the x–y plane. It is shown as OR in Fig. 1.2. Now rotation about OR is not a rotation about a coordinate axis. But this we can rectify if we first rotate the coordinate system about the z-axis (through an angle α as shown) so that, say, the y-axis becomes coincident with OR. We thus arrive at the following procedure for evaluating the effect of rotation through θ about $\hat{\mathbf{n}}$ on the coordinates (x, y, z) of any point P.

Consider the coordinates (x, y, z) of the point P in S and also (x''', y''', z''') in the frame S$'''$ obtained by rotation through θ about ON. Let these be related (in an obvious notation) via the matrix $R_{\hat{\mathbf{n}}}(\theta)$:

$$X''' = R_{\hat{\mathbf{n}}}(\theta)\, X.$$

Then, following the discussion in the previous paragraph, we can go from X to X''' by (i) rotation through α about z-axis, followed by (ii) a rotation through β about the *new* y'-axis, followed by (iii) a rotation through γ about the *new* z''-axis:

$$\begin{aligned} X''' &\equiv R_{z''}(\gamma)\, X'' \\ &\equiv R_{z''}(\gamma)\, R_{y'}(\beta)\, X' \\ &\equiv R_{z''}(\gamma)\, R_{y'}(\beta)\, R_z(\alpha) X \end{aligned}$$

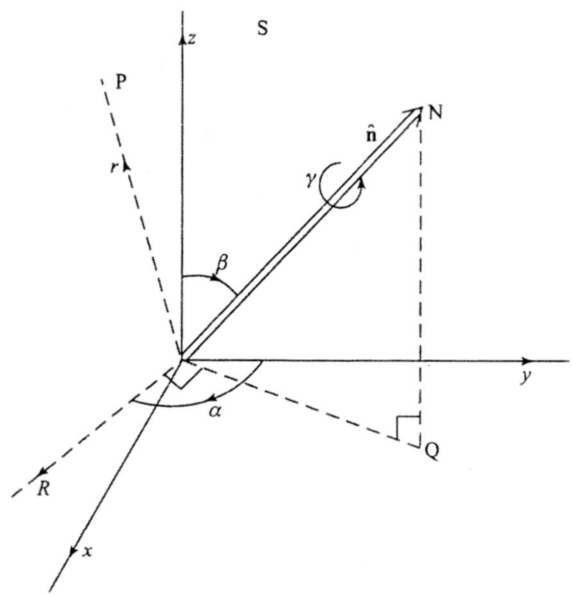

Fig. 1.2 *Euler angles*

Each of the three $R_{z''}(\gamma)$, $R_{y'}(\beta)$ $R_z(\alpha)$ about the *respective new* axes is an *elementary* rotation matrix of the form

$$\begin{bmatrix} \cos\zeta & \sin\zeta & 0 \\ -\sin\zeta & \cos\zeta & 0 \\ 0 & 0 & 1 \end{bmatrix}$$

or one obtained by exchange of rows/columns.

While the expression for $R_{\hat{n}}(\theta)$ given before has been expressed in terms of elementary rotations involving *new* y' and z'' axes, an alternative expression in terms of elementary rotations about the *original* y- and z-axes can also be derived. We state the result without proof*

$$R_{\hat{n}}(\theta) = R_z(\alpha)\, R_y(\beta)\, R_z(\gamma)$$

1.3 Definition of Vectors

An elementary definition of a vector that is usually given in order to introduce the concept is that it is a quantity that has a magnitude and a direction. For example, in Physics, a force is a vector in the sense that it has a magnitude and it acts along a definite direction. This elementary definition which has direct visual appeal may prove sufficient in elementary physics. However, normally a vector such as electric field is more conveniently described by an algebraic representation obtained by first choosing a coordinate frame and then by giving only its components $\mathbf{E} = (E_1, E_2, E_3)$, rather than by giving its magnitude and direction at each point. But these components of the same vector differ in different coordinate systems, though of course they are related to E_1, E_2, E_3, as they are components of same electric field having a definite magnitude and direction. Now suppose we are given a triplet of numbers (E_1, E_2, E_3) with reference to a frame S, and (E_1', E_2', E_3') with reference to some other frame S′ (and nothing more), the question naturally arises as to how one is to decide if they refer to the same physical quantity represented by a vector. The expected relation between (E_1', E_2', E_3') and (E_1, E_2, E_3) will give us a way of affirming just from the component description if such a triplet does represent a vector, even without going into the calculations of magnitude and direction.

The idea involved is explained with the examples given below. It involves what is called a vector field which assigns a vector to each point. (For example, an electric field in space is a vector field because at each point in space, there is an electric vector). The idea can be well illustrated by restricting oneself to two-dimensions only, as this simplifies the discussion.

Consider an ordered doublet A defined by $A = (A_1, A_2) \equiv (y, -x)$, By this, the following is meant. Choose a coordinate system S. Choose any point P and let its coordinates be (x, y). Then the quantity 'A' is represented in terms of x, y by the ordered doublet $(y, -x)$, and more importantly, this is supposed to be valid in *any* cartesian coordinate system with the same origin. We want to test if this is a vector field. This requires, in particular, that if we represent the doublet A by an arrow at P, it shall be represented by the same arrow in any frame. To test this consider two frames S and S′ as follows (Fig. 1.3).

For illustration purposes suppose S′ is obtained from S by rotation through 90°. Let the coordinates of P in S be, say, (2, 1). Then its coordinates in S′ are (1, −2). Then the doublet

*See for example *Quantum Mechanics* by A. Messiah, vol. II, John Wiley and Sons, 1966.

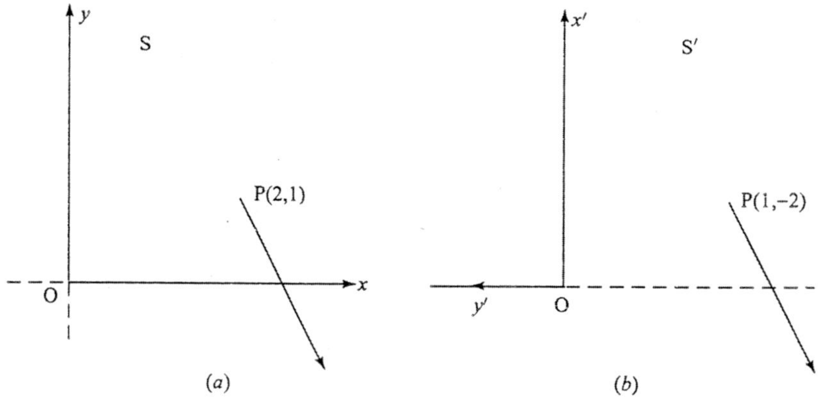

Fig. 1.3

'A' $\equiv (y, -x)$ will be represented by $(1, -2)$ in S and by $(-2, -1)$ in S′. If $(1, -2)$ is plotted as an arrow in S, it would be as shown in Fig. 1.3(a). When the doublet $(-2, -1)$ is plotted in S′ as an arrow (i.e. with x' component $= -2$; y' component $= -1$), it is seen to be given exactly by the same arrow. This suggests that the given doublet at P may be a vector, which however is confirmed only when the arrow so drawn is the same for any S′. (This indeed turns out to be the case as seen later.) In order that A is a vector field, the above must hold at each point P.

The above condition may seem trivial. To see that not any doublet represents a vector (field), consider the doublet $B = (y, 0)$. Consider the same two frames as in Fig. 1.3 for illustration and consider the same P as there. Then B will be described by doublet $(1, 0)$ at P in S and $(-2, 0)$ at P in S′. The respective arrows drawn as before in S and S′ are as given in Fig. 1.4.

The two arrows have neither the same magnitude nor the same direction. Then, $B = (y, 0)$ does not represent a vector at P and thus cannot be a vector field.

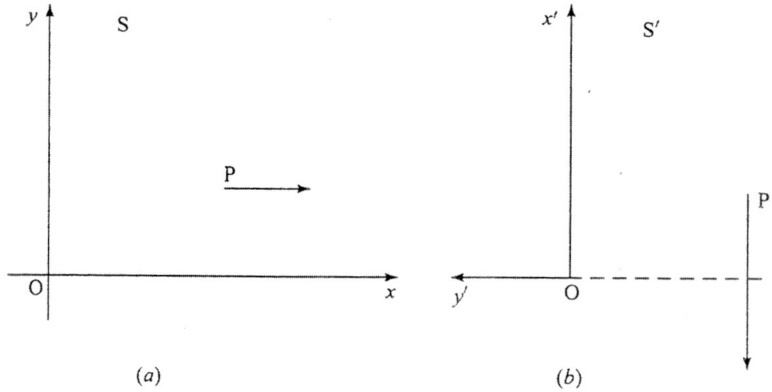

Fig. 1.4

The above way of checking whether a given entity is a vector (field) is, of course, not very suitable in practice. One would prefer an algebraic way of testing whether a quantity is a vector. This is done in the next subsection.

1.3.1 Algebraic definition of a vector

To give such a definition, we shall first find it convenient to consider the transformation laws for the components of a position vector (viz. coordinates) as it transforms from one frame to another.

Components of **r** in S′, viz. x'_1, x'_2, x'_3, are obtained from components of **r** and S, viz. x_1, x_2, x_3 by the transformation rule derived earlier in Section 1.1. It is

$$x'_i = \sum_j A_{ij}\, x_j. \tag{1.17}$$

From the above relation the transformation laws for components of a vector **V** can be easily obtained. As one knows, a given vector can be translated parallel to itself. The vector **V**, a given arrow, can be so drawn that its tail coincides with O. Let its tip be at a point P. We can then use the same diagram as Fig. 1.5 in which **r** is replaced by **V**. Then one would have exactly the same relation amongst the components of **V** as that given by Eq. (1.17) viz.

$$V'_i = \sum_j A_{ij}\, V_j. \tag{1.18}$$

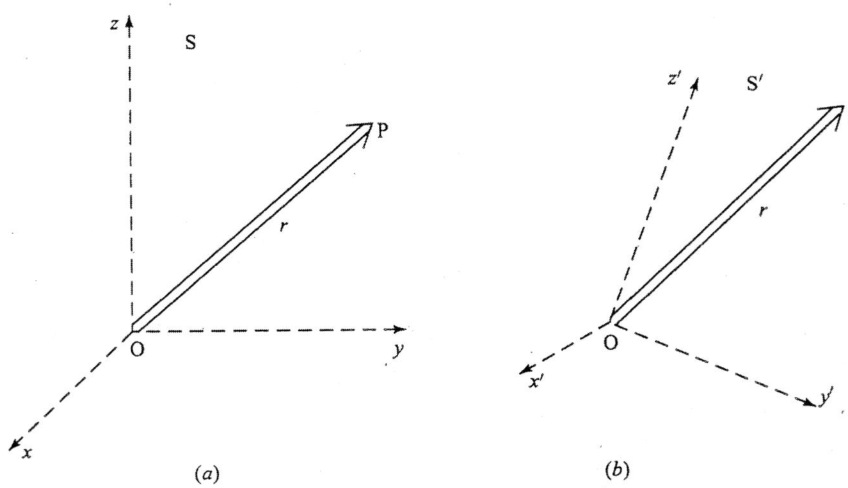

Fig. 1.5 *Transformation of components of a vector*

Here we have expanded

$$\begin{aligned} \mathbf{V} &= \sum_i V_i\, \hat{\varepsilon}_i \qquad \text{in S} \\ &= \sum_i V_i'\, \hat{\varepsilon}_i' \qquad \text{in S}'. \end{aligned} \tag{1.19}$$

We are now ready to give the algebraic definition of a vector.

Definition: An entity V represented by an ordered triplet (V_1, V_2, V_3) in S is a vector if after coordinate rotations, it is represented by (V_1', V_2', V_3') in S$'$ given by

$$V_j' = \sum A_{ij}\, V_j$$

for every cartesian coordinate system S$'$ related to S by an arbitrary coordinate rotation.

Solved example 1.1 In two-dimensions, a field is described by $\mathbf{A}(\mathbf{r}) = \left[\frac{-y}{r^2}, \frac{x}{r^2}\right]$. Show that $\mathbf{A}$ represents a vector field.

Solution: Under coordinate transformation, coordinates (x, y) transform as

$$\begin{aligned} x' &= x\,\cos\theta + y\,\sin\theta. \\ y' &= -x\,\sin\theta + y\,\cos\theta. \end{aligned} \tag{1.20}$$

$$\mathbf{A}(\mathbf{r}) = \left[-\frac{y}{r^2}, \frac{x}{r^2}\right] \qquad r^2 = x^2 + y^2. \tag{1.21}$$

$$\mathbf{A}'(\mathbf{r}') = \left[\frac{-y'}{r'^2}, \frac{x'}{r'^2}\right] = \left[\frac{-y'}{r^2}, \frac{x'}{r^2}\right]; \quad r^2 = r'^2. \tag{1.22}$$

In order that $\mathbf{A}(\mathbf{r})$ transform as a vector field, it is necessary that components of $\mathbf{A}(\mathbf{r})$ transform as

$$\begin{aligned} A_x'(\mathbf{r}') &= A_x(\mathbf{r})\,\cos\theta + A_y(\mathbf{r})\,\sin\theta, \\ A_y'(\mathbf{r}') &= -A_x(\mathbf{r})\,\sin\theta + A_y(\mathbf{r})\,\cos\theta. \end{aligned}$$

Here $\mathbf{r}$ and $\mathbf{r}'$ refer to the same physical point.

From the above,

$$A_x(\mathbf{r}) = -\frac{y}{r^2}, \quad A_y(\mathbf{r}) = \frac{x}{r^2}; \quad A_x'(\mathbf{r}') = -\frac{y'}{r^2}, \quad A_y'(\mathbf{r}') = \frac{x'}{r^2}.$$

Hence we need to prove

$$\begin{aligned} -\frac{y'}{r^2} &= \frac{-y}{r^2}\,\cos\theta + \frac{x}{r^2}\,\sin\theta, \\ \frac{x'}{r^2} &= \frac{y}{r^2}\,\sin\theta + \frac{x}{r^2}\,\cos\theta, \end{aligned} \tag{1.23}$$

which are immediately verified from (1.20).

Note: The magnetic field due to an infinite uniform current carrying wire along z-axis outside the wire is indeed proportional to the vector field considered here.

Theorem: $\nabla \equiv \sum_i \frac{\partial}{\partial x^i} \hat{\varepsilon}_i \equiv \left[\frac{\partial}{\partial x^1}, \frac{\partial}{\partial x^2}, \frac{\partial}{\partial x^3}\right]$ is a vector operator.

We know

$$x'^i = A_{ij} x^j.$$

From this it follows that

$$\frac{\partial}{\partial x^j} = \sum_i \frac{\partial x'^i}{\partial x^j} \frac{\partial}{\partial x'^i}$$
$$= \sum_i A_{ij} \frac{\partial}{\partial x'^i}$$

Multiplying the above relation by A_{kj} and summing over j, we have

$$\sum_j A_{kj} \frac{\partial}{\partial x^j} = \sum_i \left\{ \sum_j A_{ij} A_{kj} \right\} \frac{\partial}{\partial x'^i} = \sum_i \delta_{ik} \frac{\partial}{\partial x'^i} = \frac{\partial}{\partial x'^k}$$

or

$$\frac{\partial}{\partial x'^i} = \sum_j A_{ij} \frac{\partial}{\partial x^j}$$

which is identical to the transformation law (1.18) for a vector with $V_i = \frac{\partial}{\partial x^i}$.

Thus ∇, which is a differential operator and also a vector, is called a vector operator.

1.3.2 Pseudoscalars and pseudovectors

In three-dimensions, the general orthogonal coordinate transformation in a Cartesian coordinate system has been shown to be of two types: that with $\det A = +1$ and that with $\det A = -1$. The former, as we have seen, corresponds to pure coordinate rotations. The latter can always be interpreted as a combination of a pure coordinate rotation and the parity transformation. The parity transformation is a transformation of coordinate axes which changes the sense of all axes. In particular, parity takes one from a right-handed coordinate system to a left-handed coordinate system.

Thus, the behaviour of physical quantities under coordinate transformations (with origin fixed) is completely specified if the same is specified for pure rotations and for the parity transformation. We have already discussed the behaviour of physical quantities under pure rotations and classified them as scalars, vectors and, as we will learn in Section 1.4, tensors of various ranks. Now, we shall give a further classification of these by considering their behaviour under parity.

First, we shall express the parity transformation. Let (x, y, z) be coordinates of a point Q in one cartesian coordinate system S. The parity transformed system S$'$ is one whose origin is unchanged but the sense of axes is reversed. In this coordinate system, Q is represented by coordinates x', y', z' with $\mathbf{r}' = -\mathbf{r}$. Let a physical quantity (a field) in S be represented by a function $\phi(\mathbf{r})$ and let the same be described in S$'$ by $\phi'(\mathbf{r}')$. Then, at the same point Q, we compare these two:

If $\phi'(\mathbf{r}') = \phi(\mathbf{r})$	then ϕ is a scalar
$= -\phi(\mathbf{r})$	then ϕ is a pseudoscalar.

Similarly, let $\mathbf{A}(\mathbf{r})$ be a physical quantity that transforms as a vector under pure rotations and let $\mathbf{A}'(\mathbf{r}') = \sum_i A'_i(\mathbf{r}')\hat{\varepsilon}'_i$ be the same physical quantity in S$'$. Then, if for each $i = 1, 2, 3$,

$A'_i(\mathbf{r}') = -A_i(\mathbf{r})$	then A is a vector
$= +A_i(\mathbf{r})$	then A is a pseudovector.

Examples

(1) Consider a point charge q at the origin. In S, its electrostatic potential is given by

$$\phi(\mathbf{r}) = \frac{q}{4\pi\varepsilon_0}\frac{1}{\sqrt{x^2+y^2+z^2}}.$$

The same physical quantity is given in S$'$ by

$$\phi'(\mathbf{r}') = \frac{q}{4\pi\varepsilon_0}\frac{1}{\sqrt{x'^2+y'^2+z'^2}}.$$

Clearly $\phi'(\mathbf{r}') = \phi(\mathbf{r})$; and thus the above electrostatic potential is a scalar quantity. This can be verified for any arbitrary charge configuration.

(2) Consider the electric field configuration for the same point charge. In S,

$$\mathbf{E}(\mathbf{r}) = \frac{q}{4\pi\varepsilon_0}\frac{x\hat{\mathbf{i}}+y\hat{\mathbf{j}}+z\hat{\mathbf{k}}}{(x^2+y^2+z^2)^{3/2}}$$

In S$'$

$$\mathbf{E}'(\mathbf{r}') = \frac{q}{4\pi\varepsilon_0}\frac{x'\,\hat{\mathbf{i}}'+y'\,\hat{\mathbf{j}}'+z'\,\hat{\mathbf{k}}'}{(x'^2+y'^2+z'^2)^{3/2}} = \frac{q}{4\pi\varepsilon_0}\frac{-x\,\hat{\mathbf{i}}'-y\,\hat{\mathbf{j}}'-z\,\hat{\mathbf{k}}'}{(x^2+y^2+z^2)^{3/2}}.$$

Thus,

$$E'_{x'}(\mathbf{r}') = -E_x(\mathbf{r}), \quad \text{etc.}$$

This shows that $\mathbf{E}(\mathbf{r})$ is a vector field (as opposed to a pseudovector).

(3) Consider an infinite current-carrying wire (current i_o) along the z-axis. Its magnetic field in S is expressed as

$$\mathbf{B}(\mathbf{r}) = \frac{\mu_o i_o}{2\pi} \times \frac{(-y\,\hat{\mathbf{i}} + x\,\hat{\mathbf{j}})}{x^2 + y^2}.$$

As seen in S′, the wire carries a current $-i_o$ along the z'-axis. Its magnetic field is then described in S′ by

$$\mathbf{B}'(\mathbf{r}') = \frac{\mu_o(-i_o)}{2\pi} \times \frac{(-y'\,\hat{\mathbf{i}}' + x'\,\hat{\mathbf{j}}')}{x'^2 + y'^2} = \frac{\mu_o i_o}{2\pi}\frac{(-y\,\hat{\mathbf{i}}' + x\,\hat{\mathbf{j}}')}{x^2 + y^2}$$

Thus, we see that $B'_{x'} = B_x$, $B'_{y'} = B_y$, $B'_{z'} = B_z = 0$.
Thus $\mathbf{B}$ is a pseudovector.

(4) Consider the quantity $\mathbf{E}(\mathbf{r})\cdot\mathbf{B}(\mathbf{r})$. From the vector nature of $\mathbf{E}$ and pseudovector nature of $\mathbf{B}$, it is easily seen that

$$\mathbf{E}'(\mathbf{r}')\cdot\mathbf{B}'(\mathbf{r}') = -\mathbf{E}(\mathbf{r})\cdot\mathbf{B}(\mathbf{r})$$

Thus $\mathbf{E}\cdot\mathbf{B}$ is a pseudoscalar.

1.4 Cartesian Tensors

Quantities which are invariant under coordinate rotations are called scalars. There are many examples of scalars, e.g., mass of a body, electric charge of a particle, kinetic energy of a particle, etc. We have already been introduced to the definition of vectors and pseudovectors. These also occur in Physics very commonly. Velocity, acceleration, and electric field are vectors, magnetic field and angular momentum are pseudovectors. But the scalars (given by one number), the vectors (given by an ordered triplet) turn out to be only the first two members of a series of objects generally known as the Cartesian tensors, and do not exhaust quantities of physical interest. For example, the inertial properties of a right body under rotation are represented by an "inertia tensor" $I_{ij}(1 \le i, j \le 3)$ which has 9 components. This, as we shall see, is a cartesian tensor of rank two. In a similar manner, the electric permittivity of a non-isotropic material is represented by ε_{ij} a second rank tensor. The purpose of this section is to generalize the notion of a vector, and thus give the definition of cartesian tensors of arbitrary rank.

How would one generalize objects such as vectors? We saw that the transformation law for a general vector was abstracted from the coordinate transformation law. One would consider the bilinear combination of coordinates $x_i\,(i = 1, 2, 3)$, e.g. consider $x_i x_j$. This is easily seen to transform as

$$x_i' x_j' = \left[\sum_k A_{ik}\, x_k\right]\left[\sum_i A_{jl}\, x_l\right]$$
$$= \sum_k \sum_l A_{ik}\, A_{jl} (x_k\, x_l).$$

From this, we shall abstract the definition of a cartesian tensor of rank 2 given below.

Unlike a vector, the cartesian tensor T of rank two is an object with 9 components T_{ij}, where i and j take values 1, 2, 3. This entity is represented in any other Cartesian frame S′ (obtained by a coordinate rotation) by (possibly) another set of 9 numbers T'_{ij}. T is a Cartesian tensor of rank 2 iff T'_{ij} are related to T_{ij} by

$$T'_{ij} = \sum_k \sum_l A_{ik}\, A_{jl}\, T_{kl}. \tag{1.24}$$

The above definition is a natural generalization of the definition of a vector V in that the transformation matrix A is introduced for both of the two indices i, j of T.

1.4.1 Symmetric and antisymmetric tensors of rank two

A general rank-two tensor has 9 independent components. However, in Physics, certain special cases of a tensor of rank two often appear naturally.

(i) Symmetric tensor: A tensor T_{ij} is symmetric iff

$$T_{ij} = T_{ji} \qquad 1 \le i, j \le 3.$$

Such a tensor has 6 independent components. This is seen as follows: The above constraint does not restrict T_{11}, T_{22}, T_{33}; whereas the constraints results in the remaining six components to consist of these equal pairs ($T_{12} = T_{21}$, etc.). Thus three out of nine components are dependent, etc.

(ii) Antisymmetric tensor: A tensor T_{ij} is antisymmetric iff

$$T_{ij} = -T_{ji}.$$

Such a tensor has 3 independent components. The "diagonal components" T_{11}, T_{22}, T_{33} vanish; while $T_{21} = -T_{12}$, $T_{32} = -T_{23}$, $T_{13} = -T_{31}$, etc.

Every tensor of rank two can be written as a sum of a symmetric tensor and an antisymmetric tensor in a unique fashion:

$$T_{ij} = \frac{1}{2}\left[T_{ij} + T_{ji}\right] + \frac{1}{2}\left[T_{ij} - T_{ji}\right]$$
$$\equiv S_{ij} + A_{ij}.$$

As is easily seen, S_{ij} and A_{ij} are symmetric and antisymmetric tensors respectively.

Examples of symmetric and antisymmetric tensors occurring in Physics

(i) Inertia tensor is defined by

$$I_{ij} = \int d\tau\rho(\mathbf{r})\left\{-x_i\, x_j + r^2\delta_{ij}\right\}$$

is evidently symmetric: $I_{ij} = I_{ji}$.

(ii) Consider the vector operator ∇ and a vector field $\mathbf{A}(\mathbf{r})$. Consider the nine quantities $\partial_i A_j$ constituted in terms of its components. $\partial_i A_j$ transforms as a second rank tensor. Then

$$F_{ij} \equiv \partial_i A_j - \partial_j A_i$$

is an antisymmetric tensor of rank two. If we let $\mathbf{A}$ to be vector potential, the F_{ij} then are just components of $(\nabla \times \mathbf{A})$, i.e. of magnetic field vector $\mathbf{B}$. In fact,

$$B_k = \frac{1}{2}\varepsilon_{ijk}F_{ij}.$$

(Note that this antisymmetric tensor is equivalent to a vector/pseudovector depending on whether $\mathbf{A}$ is a pseudovector or a vector.)

Definition: Trace of a tensor: Let T_{ij} be a tensor of rank two. Consider

$$\mathrm{tr}T \equiv \sum_{j=i}^{3} T_{jj}.$$

This is called trace of T.

As can be easily shown (see Problem 1.21), this is invariant under rotation, i.e., a scalar.

Note that tr$T$ is a scalar constructed from a tensor. This will be seen later to be a special case of a process known as contraction of tensor.

1.4.2 Cartesian tensor of rank n

Definition: Consider a set of 3^n objects $T_{i_1\cdots i_n}$ in frame S $(1 \le i_i, \cdots, i_n \le 3)$. This set is called a tensor of rank n iff it transforms under coordinate rotations according to the law

$$T'_{j_1\cdots j_n} = \sum_{i_1\cdots i_n} A_{j_1 i_1} \cdots A_{j_n i_n} T_{i_1\cdots i_n}.$$

The above definition is an obvious generalization of definition of a tensor of rank two.

Examples $x_i x_j x_k$, $\delta_{ij} x_m x_n$ are tensors of rank three and four respectively.

1.4.2.1 Pseudotensors

When we discussed scalars and vectors, we noted that each of these came in two types (say scalars or pseudoscalars) which are distinguished from each other by their behaviour under improper rotations. Thus, the general definition of a scalar field and a pseudoscalar field can be written as

$$\begin{aligned} \varphi'_s(\mathbf{r}') &= \varphi'_s(\mathbf{r}) && \text{scalar} \\ \varphi'_p(\mathbf{r}') &= \det A \varphi'_p(\mathbf{r}) && \text{pseudoscalar} \end{aligned} \tag{1.25}$$

for *any* rotation A, proper or improper. Similarly, the definitions of vectors and pseudovector fields is

$$\begin{aligned} V'_i(\mathbf{r}') &= A_{ij} V_j(\mathbf{r}) && \text{vector} \\ P'_i(\mathbf{r}') &= (\det A) A_{ij}\ P_j(\mathbf{r}') && \text{pseudovector.} \end{aligned} \tag{1.26}$$

It is then natural to generalize the above definitions to tensors of higher rank.

Thus a pseudotensor of rank n would instead satisfy:

$$T'_{j_1\cdots j_n} = (\det A) A_{j_1 i_1} \cdots A_{j_n i_n} T_{i_1\cdots i_n} \tag{1.27}$$

for an arbitrary rotation A, proper or improper.

1.4.3 Special tensors

We shall now introduce certain useful tensors known as the Kronecker Delta (δ_{ij}) and the Levi–Civita tensor (ε_{ijk}).

Kronecker delta δ_{ij} has already been defined earlier as

$$\begin{aligned} \delta_{ij} &= 1 && i = j \\ &= 0 && i \ne j && i, j = 1, 2, 3 \end{aligned} \tag{1.28}$$

δ_{ij} is an invariant tensor. By this we mean the following: Suppose we define the Kronecker delta components to have values δ_{ij} in every frame. Then the object so defined transforms as a tensor of rank 2 under rotation. To see this, we must have the law of (1.24) verified if we substitute $T'_{ij} = T_{ij}$, i.e., we must have

$$\delta_{ij} = \sum_k \sum_l A_{ik} A_{jl} \delta_{kl}. \tag{1.29}$$

But the R.H.S. is, from the orthogonal nature of matrix A,

$$= \sum_k A_{ik} A_{jk} = \delta_{ij}.$$

Thus (1.29) is immediately verified.

Unlike many other tensors/vectors, the components of Kronecker delta have the same values in all Cartesian frames. Hence it is called an invariant tensor. But unlike a scalar it has 9 components.

Levi–Civita tensor This is a pseudotensor ε_{ijk} of rank 3, defined uniquely by:

(a) ε_{ijk} is completely antisymmetric,
(b) $\varepsilon_{123} = 1$.

By complete antisymmetry of ε_{ijk} we mean that it changes sign if any pair of indices are interchanged. Thus, e.g., $\varepsilon_{213} = -\varepsilon_{123} = -1$. In particular, when any two indices coincide, that particular ε symbol must vanish (e.g. $\varepsilon_{113} = -\varepsilon_{113} = 0$). From the above defining relations, ε_{ijk} is uniquely defined; with the only nonzero components given by

$$\varepsilon_{123} = \varepsilon_{231} = \varepsilon_{312} = -\varepsilon_{132} = -\varepsilon_{321} = -\varepsilon_{213} = 1. \tag{1.30}$$

Let us now show that ε_{ijk} transforms like an invariant pseudotensor of rank three under rotations of coordinate axes. Thus we have to verify that

$$T'_{ijk} = \det A \sum_l \sum_m \sum_n A_{il} A_{jm} A_{kn} T_{lmn} \tag{1.31}$$

with $T_{ijk} = \varepsilon_{ijk} = T'_{ijk}$, since an invariant tensor must have same matrix elements in both frames. The R.H.S. of (1.31) is

$$\det A \sum_l \sum_m \sum_n \varepsilon_{lmn} A_{il} A_{jm} A_{kn}. \tag{1.32}$$

For $i = 1, j = 2, k = 3$, this is just the expansion of $\det A$. Moreover, the above object is antisymmetric under any pair of indices (ijk). Hence the above quantity (1.32) is

$$\det A \varepsilon_{ijk} \det A = \varepsilon_{ijk} = \text{L.H.S. of (1.31)}$$

as $(\det A)^2$ is always 1. This proves the assumption that ε_{ijk} is an invariant pseudotensor of rank three.

1.4.4 Construction of new tensors out of given tensors

(1) *Tensor product:* Let $T^{(1)}_{i_1\cdots i_n}$ and $T^{(2)}_{j_1\cdots j_n}$ be the two tensors of ranks n and m respectively. Consider the set of $3^n \times 3^m (= 3^{n+m})$ objects

$$T^{(3)}_{k_1\cdots k_n\cdots k_{n+m}} = T^{(1)}_{k_1\cdots k_n}\, T^{(2)}_{k_{n+1}\cdots k_{n+m}} \quad 1 \le k_i \le 3 \quad \text{for} \quad 1 \le i \le n+m. \tag{1.33}$$

It is easy to see that $T^{(3)}$ has the transformation property of a tensor of rank $(n+m)$ (Verify!) Thus, we have, via "tensor product" obtained a new tensor of rank $(n+m)$ out of two tensors $T^{(1)}$ and $T^{(2)}$ of rank n and m respectively. We shall denote $T^{(3)}$ as the direct product of $T^{(1)}$ and $T^{(2)}$, viz.,

$$T^{(3)} = T^{(1)} \otimes T^{(2)}. \tag{1.34}$$

Eq. (1.34) represents, in a compact form, exactly what (1.33) represents explicitly.

(2) *Contraction of a tensor:* Consider a tensor of rank n : $T_{i_1\cdots i_n}$. Consider any two of its indices, say pth and qth $(p < q)$. Set them equal and sum over this index from 1 to 3. We then obtain a quantity with $n-2$ indices which we denote by

$$(T')_{i_1\cdots i_{p-1}\, i_{p+1}\cdots i_{q-1}\, i_{q+1}\cdots i_n} \equiv \sum_{j=1}^{3} T_{i_1\cdots i_{p-1}\; j\; i_{p+1}\cdots i_{q-1}\; j\; i_{q+1}\; i_n}.$$

In Section 1.4.1 we proved that for a second rank tensor T, tr$T$ is a scalar. Now we can follow the same proof to show that T' given above is a tensor or rank $n-2$. (This is left as an exercise.) Thus one has obtained a tensor of rank $(n-2)$ from a tensor of rank n by what is called a contraction. The process can, further, be repeated. From T' one can, by a further contraction, construct a tensor T'' of rank $n-4$, etc.

(3) *Contraction of two tensors:* Consider two tensors $T^{(1)}$ and $T^{(2)}$ of rank n and m respectively. One can contract any one index of T with any one index of T'. Thus we can construct

$$T^{(3)}_{i_1\cdots \hat{\mathbf{i}}_p\cdots i_n\; j_1\cdots \hat{\mathbf{j}}_q\cdots j_n} = \sum_{k=1}^{3} T^{(1)}_{i_1\cdots i_{p-1}\; k\cdots i_n}\, T^{(2)}_{j_1\cdots j_{q-1}\; k\cdots j_n}.$$

This, of course, can be looked upon as a contraction of tensor $T^{(1)} \otimes T^{(2)}$ constructed earlier. (In the above equation ^ denotes a missing index.)

Such contractions can be be carried out for more pairs of indices yielding successively tensors of ranks $n+m-2$, $n+m-4$, etc.

Example $\varepsilon_{ijk} F_{jk} = B_i$ is a tensor of rank $3+2-4 = 1$; i.e., a vector (see Problem 1.22).

Solved example 1.2 Consider $T_{ij} \equiv x_i x_j - \delta_{ij} r^2$. (i) Evaluate $\partial_i T_{ij}$ and T_{ii} explicitly. (ii) Write down the rank of tensors $\partial_i T_{ij}$, T_{ii} and $\varepsilon_{ijk} T_{ij}$.

Solution:

(i) $T_{ij} = x_i x_j - \delta_{ij} r^2$.

(a)
$$\partial_i T_{ij} = \partial_i (x_i x_j) - \delta_{ij} \partial_i (r^2)$$
$$= \partial_i (x_i) x_j + x_i \partial_i x_j - \delta_{ij} (2x_i).$$

We now use $\partial_i x_i = 3$ and $\partial_i x_j = \delta_{ij}$ to find

$$\partial_i T_{ij} = 3x_j + x_i \delta_{ij} - \delta_{ij} 2x_i$$
$$= 3x_j + x_j - 2x_j = 2x_j.$$

(b) $T_{ii} = x_i x_i - \delta_{ii} r^2$.

We use, $\delta_{ii} = 3$, $x_i x_i = r^2$

$$T_{ii} = r^2 - 3r^2 = -2r^2.$$

(ii) From the expression for $\partial_i T_{ij}$ and T_{ii}, it is clear that $\partial_i T_{ij} = 2x_j$ are components of a vector $2\mathbf{r}$ and $T_{ii} = -2r^2$ is a scalar.
This is also evident from the general results of Section 1.4.4. $\partial_k T_{ij}$ is a tensor of rank $1 + 2 = 3$. $\partial_i T_{ij}$ is obtained by contraction of a pair of indices. Hence its rank is $3 - 2 = 1$. Hence it is a vector.

In a similar manner, rank of $T_{ii} = 2 - 2 = 0$. Rank of $\varepsilon_{ijk} T_{ij}$ is obtained by first adding ranks of the two tensors ε and T and then subtracting two for every pair of contracted indices. Hence it is $3 + 2 - 2(2) = 1$.

Note that in each case the rank equals the number of free indices.

1.4.5 Algebra of Levi–Civita tensor and Kronecker delta

As below, we introduce unit vectors $\hat{\mathbf{e}}_i$ of a right handed cartesian coordinate system. They satisfy

$$\hat{\mathbf{e}}_i \cdot \hat{\mathbf{e}}_j = \delta_{ij} \qquad i, j = 1, 2, 3 \tag{1.35}$$

and

$$\hat{\mathbf{e}}_i \times \hat{\mathbf{e}}_j = \varepsilon_{ijk} \hat{\mathbf{e}}_k \qquad i, j = 1, 2, 3. \tag{1.36}$$

The following properties are easily verified and are left as exercises. (Note the use of the summation convention.)

(i) $\delta_{ij} C_j = C_i$

(ii) $\delta_{ii} = 3$

(iii) $\varepsilon_{iik} = 0$

(iv) $\delta_{ij} \varepsilon_{jkl} = \varepsilon_{ikl}$

(v) $\delta_{ij} A_i B_j = A \cdot B$

(vi) $\varepsilon_{ijk} \hat{\mathbf{e}}_i A_j B_k = \mathbf{A} \times \mathbf{B}$ (1.37)

We shall now prove further properties of ε and δ.

(vii) $\varepsilon_{ijk} \varepsilon_{ijl} = 2\delta_{kl}$ (1.38)

For a given k and l, the L.H.S. has 9 terms since i and j can take 3 values each, independently. In order that any of such terms is nonzero, i and j must be distinct and k must be distinct from both i and j and so must be equal to l. This can only happen if k and l are identical. (Verify!) Hence, L.H.S. is zero unless $k = l$. When $k = l$ (dropping the summation convention).

$$\text{L.H.S.} = \sum_{i,j} \varepsilon_{ijk}\, \varepsilon_{ijk} = \sum_{i,j} \varepsilon_{ijk}^2.$$

Now for a given k, there are only two pairs of (i, j) distinct from k for which $\varepsilon_{ijk} \neq 0$, and $\varepsilon_{ijk}^2 = 1$ in both of these cases.

$$\text{Hence L.H.S.} = 2 \text{ if } k = l.$$

It thus follows that

$$\text{L.H.S.} = 2\delta_{kl} = \text{R.H.S.}$$

(viii) $\varepsilon_{ijk} S_{ij} = 0$ if $S_{ij} = S_{ji}$, i.e. S is a symmetric tensor.

$$\text{L.H.S.} = \varepsilon_{ijk} S_{ij} = \varepsilon_{ijk} S_{ji} \qquad (S_{ji} = S_{ij})$$
$$= -\varepsilon_{jik} S_{ji}.$$

Since i and j are dummy (summation) indices they can be replaced by j and i respectively. Hence

$$\text{L.H.S.} = -\varepsilon_{ijk} S_{ij} = -\text{L.H.S.}$$

Hence

$$\text{L.H.S.} = \varepsilon_{ijk} S_{ij} = 0. \tag{1.39}$$

(ix) $\varepsilon_{ijk}\,\varepsilon_{ilm} = \delta_{jl}\,\delta_{km} - \delta_{jm}\,\delta_{kl}$ (1.40)

Here, j, k, l and m are free indices and i is the summation index going from 1 to 3. Hence for a given j, k, l and m, there are three terms on the left-hand side. Now for a term with given values of i, the two ε symbols on the L.H.S. will both be nonzero only when j and k are distinct and both not equal to i, and further, only when l and m are distinct and both not equal to i. Now, since, any free index takes only three values, this requires that j is either equal to l or m and k is equal to the remaining of l and m. Otherwise the term under consideration is zero.

Here the L.H.S. is of the form

$$\text{L.H.S.} = A\delta_{jl}\,\delta_{km} + B\delta_{jm}\,\delta_{kl}. \tag{1.41}$$

Now the left-hand side is antisymmetric in j and k (i.e., changes sign under the interchange of j and k). The R.H.S. of (1.41) will also be so iff $A = -B$. Hence

$$\text{L.H.S.} = A(\delta_{jl}\,\delta_{km} - \delta_{jm}\,\delta_{kl}). \tag{1.42}$$

It may, incidently, be noticed that now both L.H.S. and R.H.S. of (1.42) are antisymmetric in (l, m) also. To determine A, we choose $j = l = 2, k = m = 3$. Here only one term contributes to the L.H.S. owing to the antisymmetry of ε symbol, and it is with $i = 1$. This term is $\varepsilon_{123}\varepsilon_{123} = 1$. A comparison of both sides now yields

$$1 = A\delta_{22}\,\delta_{33} = A.$$

Hence, we have the result

$$\varepsilon_{ijk}\,\varepsilon_{ilm} = \delta_{jl}\,\delta_{km} - \delta_{jm}\,\delta_{kl}. \tag{1.43}$$

1.4.6 Applications of the use of δ_{ij} and ε_{ijk} to vector algebra/calculus

We shall now demonstrate how the use of δ_{ij} and ε_{ijk} facilitates the derivations of vector identities greatly. Students should master this technique well, so much so that, it becomes unnecessary to memorize the vector identities as they can be reproduced on the spot easily with these techniques.

(i) Consider $(\mathbf{A} \times \mathbf{B})\cdot(\mathbf{C} \times \mathbf{D})$.

$$(\mathbf{A} \times \mathbf{B}) = \varepsilon_{ijk}\,\hat{\mathbf{e}}_i\,A_j B_k$$

$$(\mathbf{C} \times \mathbf{D}) = \varepsilon_{lmn}\,\hat{\mathbf{e}}_l\,C_m D_n$$

We take the dot product of these and note $\hat{\mathbf{e}}_i \cdot \hat{\mathbf{e}}_l = \delta_{il}$ to obtain

$$\begin{aligned}
(\mathbf{A} \times \mathbf{B}) \cdot (\mathbf{C} \times \mathbf{D}) &= \varepsilon_{ijk} A_j B_k \delta_{il} \varepsilon_{lmn} C_m D_n \\
&= \varepsilon_{ijk} A_j B_k \varepsilon_{imn} C_m D_n \\
&= \varepsilon_{ijk}\, \varepsilon_{imn} A_j B_k C_m D_n \\
&= (\delta_{jm}\delta_{kn} - \delta_{jn}\delta_{km}) A_j B_k C_m D_n \\
&= [(\delta_{jm} A_j C_m)(\delta_{kn} B_k D_n) - (\delta_{jn} A_j D_n)(\delta_{km} B_k C_m)] \\
&= \mathbf{A}\cdot\mathbf{C}\,\mathbf{B}\cdot\mathbf{D} - \mathbf{A}\cdot\mathbf{D}\,\mathbf{B}\cdot\mathbf{C} \qquad (1.44)
\end{aligned}$$

In a very similar manner, one can show that (see Problem 1.24)

$$\mathbf{A} \times (\mathbf{B} \times \mathbf{C}) = \mathbf{A}\cdot\mathbf{C}\mathbf{B} - (\mathbf{A}\cdot\mathbf{B})\mathbf{C} \qquad (1.45)$$

We now express various "derivatives" of scalars and vectors using $\nabla = \hat{\mathbf{e}}_i \partial_i$.

$$\begin{aligned}
&\textit{Gradient} &&: \nabla\phi &&= \hat{\mathbf{e}}_i \partial_i \phi \\
&\textit{Divergence} &&: \nabla\cdot\mathbf{A} &&= (\hat{\mathbf{e}}_i \partial_i)\cdot(\hat{\mathbf{e}}_j A_j) = \delta_{ij}\partial_i A_j = \partial_i A_i \\
&\textit{Curl} &&: \nabla\times\mathbf{A} &&= (\hat{\mathbf{e}}_i \partial_i) \times (\hat{\mathbf{e}}_j A_j) \\
& && &&= (\hat{\mathbf{e}}_i \times \hat{\mathbf{e}}_j)\partial_i A_j = \varepsilon_{ijk}\hat{\mathbf{e}}_k \partial_i A_j \\
& && &&= \varepsilon_{ijk}\hat{\mathbf{e}}_i \partial_j A_k \quad \text{(cyclic symmetry of symbol)} \\
&\textit{Laplacian} &&: \nabla\cdot\nabla &&= \nabla^2 = (\hat{\mathbf{e}}_i \partial_i)\cdot(\hat{\mathbf{e}}_j \partial_j) = \delta_{ij}\partial_i \partial_j = \partial_i \partial_i
\end{aligned}$$

We shall, now, prove a number of vector calculus identities:

(i)

$$\begin{aligned}
\nabla\cdot(\phi\mathbf{A}) &= (\hat{\mathbf{e}}_i \partial_i)\cdot(\phi\hat{\mathbf{e}}_j A_j) \\
&= (\hat{\mathbf{e}}_i \cdot \hat{\mathbf{e}}_j)\partial_i(\phi A_j) = \delta_{ij}\partial_i(\phi A_j) \\
\partial_i(\phi A_i) &= (\partial_i \phi) A_i + \phi\partial_i A_i = \nabla\phi\cdot\mathbf{A} + \phi\nabla\cdot\mathbf{A} \qquad (1.46)
\end{aligned}$$

(ii) In a similar manner, one can show that (see Problem 1.27)

$$\nabla \times (\phi\mathbf{A}) = \nabla\phi \times \mathbf{A} + \phi\nabla \times \mathbf{A} \qquad (1.47)$$

(iii)

$$\begin{aligned}
\nabla\cdot(\mathbf{A}\times\mathbf{B}) &= (\hat{\mathbf{e}}_i \partial_i)\cdot(\hat{\mathbf{e}}_j \varepsilon_{jkl} A_k B_l) \\
&= \hat{\mathbf{e}}_i \cdot \hat{\mathbf{e}}_j \varepsilon_{jkl} \partial_i (A_k B_l) \\
&= \varepsilon_{ikl}(\partial_i A_k B_l + A_k \partial_i B_l) \\
&= (\nabla \times A)_l B_l - A_k (\nabla \times B)_k \\
&= (\nabla \times \mathbf{A})\cdot\mathbf{B} - \mathbf{A}\cdot(\nabla \times \mathbf{B}) \qquad (1.48)
\end{aligned}$$

Here we have used $(\nabla \times A)_l = \varepsilon_{lik}\partial_i A_k = \varepsilon_{ikl}\partial_i A_k$ and the antisymmetry of the symbol.

(iv) Now we come to the relatively difficult to prove identity:

$$\begin{aligned}\nabla(\mathbf{A}\cdot\mathbf{B}) &= (\hat{\mathbf{e}}_i\partial_i)(A_jB_j)\\ &= \hat{\mathbf{e}}_i\partial_i(A_jB_j)\\ &= \hat{\mathbf{e}}_i(\partial_i A_jB_j + A_j\partial_i B_j)\\ &= \hat{\mathbf{e}}_i[(\partial_i A_j - \partial_j A_i)B_j + A_j(\partial_i B_j - \partial_j B_i) + \partial_j A_i B_j + A_j\partial_j B_i]\end{aligned}$$

$\partial_i A_j - \partial_j A_i$ is a component of $(\nabla\times\mathbf{A})$. We can write $\partial_i A_j - \partial_j A_i = \varepsilon_{ijk}(\nabla\times\mathbf{A})_k$. Hence

$$\begin{aligned}\nabla(\mathbf{A}\cdot\mathbf{B}) = {} & \varepsilon_{ijk}\hat{\mathbf{e}}_j B_j(\nabla\times\mathbf{A})_k + \varepsilon_{ijk}\hat{\mathbf{e}}_i A_j(\nabla\times\mathbf{B})_k\\ & + (B_j\partial_j)A_i\hat{\mathbf{e}}_i + (A_j\partial_j)(B_i\hat{\mathbf{e}}_i).\end{aligned}$$

Noting that $\varepsilon_{ijk}\hat{\mathbf{e}}_i B_j C_k = (\mathbf{B}\times\mathbf{C})_i$ and $B_j\partial_j = \mathbf{B}\cdot\nabla$, etc., we get

$$\nabla(\mathbf{A}\cdot\mathbf{B}) = \mathbf{B}\times(\nabla\times\mathbf{A}) + \mathbf{A}\times(\nabla\times\mathbf{B}) + B\cdot\nabla\mathbf{A} + A\cdot\nabla\mathbf{B} \tag{1.49}$$

(v) $\nabla\times(\mathbf{A}\times\mathbf{B}) = \mathbf{B}\cdot\nabla\mathbf{A} - \mathbf{A}\cdot\nabla\mathbf{B} - \mathbf{B}(\nabla\cdot\mathbf{A}) + \mathbf{A}\nabla\cdot\mathbf{B}$

This identity is proven by using the two-ε identity (1.40). We leave the derivation as an exercise (see Problem 1.28).

(vi) Now we prove the commonly occurring vector identity (in the derivation of wave equation, etc.).

Consider

$$\nabla\times(\nabla\times\mathbf{A}) = \nabla\nabla\cdot\mathbf{A} - \nabla^2\mathbf{A} \tag{1.50}$$

$$\begin{aligned}\nabla\times(\nabla\times\mathbf{A}) &= \varepsilon_{ijk}\hat{\mathbf{e}}_i\partial_j(\nabla\times\mathbf{A})_k\\ &= \varepsilon_{ijk}\hat{\mathbf{e}}_i\partial_j\varepsilon_{klm}\partial_l A_m\\ &= \varepsilon_{kij}\varepsilon_{klm}\hat{\mathbf{e}}_i\partial_j\partial_l A_m.\end{aligned}$$

Here, we have used the cyclic symmetry of ε symbol. Now using the two-ε identity (1.42), this simplifies to

$$\begin{aligned}&= (\delta_{il}\delta_{jm} - \delta_{im}\delta_{jl})\hat{\mathbf{e}}_i\partial_j\partial_l A_m\\ &= \hat{\mathbf{e}}_i\partial_m\partial_i A_m - \hat{\mathbf{e}}_i\partial_j\partial_j A_i\\ &= (\hat{\mathbf{e}}_i\partial_i)(\partial_m A_m) - \partial_j\partial_j(\hat{\mathbf{e}}_i A_i)\\ &= \nabla(\nabla\cdot\mathbf{A}) - \nabla^2\mathbf{A}.\end{aligned}$$

Solved example 1.3 Magnetic induction **B** exerts a force on a magnetic moment **m** given by

$$\mathbf{F} = \nabla \times (\mathbf{B} \times \mathbf{m})$$

Show that $\mathbf{F} = \nabla(\mathbf{m}\cdot\mathbf{B})$. Here **m** is a constant vector and **B** satisfies $\nabla\cdot\mathbf{B} = 0$ and $\nabla \times \mathbf{B} = 0$. Solve directly using the ε symbols.

Solution:

$$\begin{aligned}
\nabla \times (\mathbf{B} \times \mathbf{m}) &= \varepsilon_{ijk}\hat{\mathbf{e}}_i \partial_j (\mathbf{B} \times \mathbf{m})_k \\
&= \varepsilon_{ijk}\hat{\mathbf{e}}_i \partial_j (\varepsilon_{kln} B_l m_n) \\
&= \varepsilon_{kij}\varepsilon_{kln}\hat{\mathbf{e}}_i m_n \partial_j B_l \qquad (m_n = \text{const.}) \\
&= (\delta_{il}\delta_{jn} - \delta_{in}\delta_{jl})\hat{\mathbf{e}}_i m_k \partial_j B_l \\
&= \hat{\mathbf{e}}_i m_j \partial_j B_i - \hat{\mathbf{e}}_i m_i \partial_j B_j
\end{aligned}$$

Now $\partial_j B_j = \nabla\cdot\mathbf{B} = 0$. And since $\nabla \times \mathbf{B} = 0$, $\partial_j B_i = \partial_i B_j$.

$$\begin{aligned}
\nabla \times (\mathbf{B} \times \mathbf{m}) &= \hat{\mathbf{e}}_i m_j \partial_i B_j = \hat{\mathbf{e}}_i \partial_i (m_j B_j) \\
&= \nabla(\mathbf{m}\cdot\mathbf{B})
\end{aligned}$$

Problems

On § 1.1

1.1 Verify equation (1.7) of § 1.1.

1.2 Show for $A(\theta)$ defined in § 1.1 that $\det A(\theta) = 1$ and $A^{-1}(\theta) = A(-\theta)$. What are the physical interpretations of these results?
(Hint: $dx'dy' = |\det A|dxdy$).

1.3 Verify the property (f) of § 1.1.

1.4 For an arbitrary real 2×2 matrix $A = \begin{bmatrix} a_{11} & a_{12} \\ a_{21} & a_{22} \end{bmatrix}$, write down the conditions implied by the orthogonality of A, viz. $A^{\mathrm{T}}A = 1$. From this, show that A can be written in the form of $A(\theta)$ (see (1.5) of § 1.1) provided that $\det A = 1$. In the case $\det A = -1$, show that A can be cast in the form

$$A = \begin{bmatrix} 1 & 0 \\ 0 & -1 \end{bmatrix} A(\theta)$$

Interpret the above result.

1.5 Show that the coordinate transformation equations (1.3a) and (1.3b) can be written as

$$z' = e^{i\theta} z, \qquad \text{where } z = x + iy;\ z' = x' + iy'.$$

Use this relation to verify the property (f) of § 1.1 in an easier manner.

On § 1.2

1.6 Show how you would determine a proper rotation matrix $A_{3\times 3}$ given a_{11}, a_{12} and a_{21}. Is A uniquely determined? (Note: det $A = +1$ here).

1.7 Verify that the Euler representation

$$R_{\hat{\mathbf{n}}}(\theta) = R_{z''}(\gamma)R_{y'}(\beta)R_z(\alpha)$$

satisfies (i) $R^T R = 1$, (ii) det $R = 1$.

1.8 Verify $R_{x'}(\pi)R_y(\pi) = R_{y'}(\pi)R_x(\pi)$. Thus in exceptional cases, finite rotations about two different axes do commute. This is so here as each is a diagonal matrix.

1.9 Write down the matrices for $R_x(\pi)$ and $R_y(\pi/2)$. Verify the inequality

$$R_{y'}(\pi/2)R_x(\pi) \neq R_{x'}(\pi)R_y(\pi/2)$$

by explicit evaluation of both sides.

1.10 Cast $R_x(\pi/2)$ in the standard Euler angle form given in the text. Hence determine α, β, γ.

1.11 For $\hat{\mathbf{n}} = \left[\frac{\hat{\mathrm{i}}+\hat{\mathrm{j}}+\hat{\mathrm{k}}}{\sqrt{3}}\right]$ and $\theta = \pi/2$, determine the Euler form for $R_{\hat{\mathbf{n}}}(\theta)$. Evaluate $\mathrm{tr}R_{\hat{\mathbf{n}}}(\theta)$.

1.12 Verify the Euler angle representation

$$R_x(\beta) = R_{z''}(\pi/2)R_y(-\beta)R_x(-\pi/2)$$

by applying both sides to a coordinate system.

On § 1.3

1.13 Describe which of the following constitutes a vector giving argument in each case.

i) $A_x = x \quad A_y = y \quad A_z = 0$

ii) $A_x = z \quad A_y = x \quad A_z = y$

iii) $A_x = -\frac{y}{r^3} \quad A_y = \frac{x}{r^3} \quad A_z = 0$

iv) $A_x = yz \quad A_y = xz \quad A_z = xy$

v) $A_x = \mathbf{r}\cdot\mathbf{B}x \quad A_y = \mathbf{r}\cdot\mathbf{B}y \quad A_z = \mathbf{r}\cdot\mathbf{B}z$

where B is a vector.

1.14 Prove that the triple product of vectors **A**, **B**, **C**, i.e. $\mathbf{A}\cdot\mathbf{B}\times\mathbf{C}$ is a pseudoscalar.

1.15 Prove that for two vectors (or two pseudovectors) **A** and **B**, $\mathbf{A}\times\mathbf{B}$ is a pseudovector.

1.16 Given that Classical Electrodynamics is invariant under parity operation, construct physical arguments to prove why **E**, **S** (Poyntings vector) must be vectors; **B** must be a pseudovector and electrostatic potential a scalar. Given further that the energy density u must be a bilinear in fields **E** and **B**, write down the general form for u.

On § 1.4

1.17 Given the components of three vectors **A**, **B** and **C**, construct two totally symmetric tensors of rank three and a totally antisymmetric tensor of rank three. (T_{ijk} is said to be totally symmetric/(antisymmetric) if it is unchanged/(changes sign) under the interchange of every pair of indices.

1.18 Given the tensor $T_{ijk} = x_i x_j x_k$, construct a "traceless" symmetric tensor $\tilde{T}_{ijk}$ from it such that $\tilde{T}$ is traceless with respect to every pair of indices ($\tilde{T}_{ijj} = 0 = \tilde{T}_{iik} = \tilde{T}_{iji}$). Do so by substracting terms proportional to δ_{pq} from T.

1.19 Given a tensor T_{ij}, resolve T into sum of three parts:

1. A symmetric traceless part
2. An antisymmetric part
3. A part proportional to δ_{ij}

Apply this to (i) $T_{ij} = x_i x_j$ (ii) $T_{ij} = A_i B_j$, where **A** and **B** are vectors.

1.20 Given a tensor T_{ijk} (and invariant tensors δ_{ij} and ε_{ijk}) construct as many scalars, vectors and tensors of rank two out of these which are linear or bilinear in T.

1.21 Show that for a tensor T_{ij}, the trace is a scalar/pseudoscalar depending on whether T is tensor/pseudotensor.

1.22 Verify that $B_i = \varepsilon_{ijk} F_{ij}$ (with F_{ij} a tensor of rank two) is a vector under proper rotations.

On § 1.5

1.23 Show only using δ and ε symbols that $\nabla \cdot \mathbf{r} = 3$ and $\nabla \times \mathbf{r} = 0$.

1.24 Show that $\mathbf{A} \times (\mathbf{B} \times \mathbf{C}) = \mathbf{A} \cdot \mathbf{C} \mathbf{B} - \mathbf{A} \cdot \mathbf{B} \mathbf{C}$

1.25 Show $(\mathbf{A} \times \mathbf{B}) \times (\mathbf{C} \times \mathbf{D}) = (\mathbf{A} \cdot \mathbf{B} \times \mathbf{D})\mathbf{C} - (\mathbf{A} \cdot \mathbf{B} \times \mathbf{C})\mathbf{D}$

1.26 Show that $\nabla(\mathbf{A} \cdot \mathbf{B} \times \mathbf{r}) = \mathbf{A} \times \mathbf{B}$ if **A** and **B** are constant vectors.

1.27 Show that $\nabla \times (\phi \mathbf{A}) = \nabla \phi \times \mathbf{A} + \phi \nabla \times \mathbf{A}$

1.28 Show that $\nabla \times (\mathbf{A} \times \mathbf{B}) = \mathbf{B} \cdot \nabla \mathbf{A} - \mathbf{A} \cdot \nabla \mathbf{B} - \mathbf{B}(\nabla \cdot \mathbf{A}) + \mathbf{A} \nabla \cdot \mathbf{B}$

1.29 Using the result of (1.27) calculate $\nabla \times (f(r)\mathbf{r})$; where $f(r)$ is a function of radial distance r only. Show that the result can be understood as a special case of $\nabla \times \nabla \psi(\mathbf{r}) = 0$.

1.30 Show that $\nabla \cdot (\nabla \varphi \times \nabla \psi) = 0$.

1.31 Show that every solution of $\nabla \times (\nabla \times \mathbf{A}) = k^2 \mathbf{A}$ automatically satisfies $\nabla^2 \mathbf{A} + k^2 \mathbf{A} = 0$ and $\nabla \cdot \mathbf{A} = 0$. (Hint: Take divergence of the original equation. $k^2 \neq 0$.)

1.32 Are $\mathbf{A} \times (\nabla \times \mathbf{B})$ and $(\mathbf{A} \times \nabla) \times \mathbf{B}$ the same? If not obtain a relation involving both.

2 Lorentz Transformation, 4-Vectors and 4-Tensors

This chapter will be devoted to the study of Lorentz transformations and to vectors and tensors in a 4-dimensional space–time called the Minkowski space-time. These are relevant to the study of Special Theory of Relativity which enters into the very foundation of the correct formulation of Classical Mechanics and Classical Electrodynamics, and is essential to the study of High Energy Physics. Space–time has four dimensions, three space and one time, and is broadly speaking a generalization of the usual 3-dimensional Euclidean space dealt with in the first chapter. It, however, differs in certain mathematical details and the purpose of this chapter is (partly) to emphasize these differences but for which the discussion here would have been a straightforward generalization.

2.1 General Lorentz Transformations and the Metric

In the three-dimensional space, the rotations leave the length of a vector unchanged. For example, if one considers the position vector $\mathbf{r}$ of a point P given by $\mathbf{r} \equiv (x_1, x_2, x_3)$, then $\sum_{i=1}^{3} x_i^2$ is invariant under rotations about the origin. The four-dimensional space–time (the Minkowski space–time) differs from this in a mathematical sense in that the obvious generalization $x_1^2+x_2^2+x_3^2+x_0^2$ is not invariant (here $x_0 = ct$) under a Lorentz transformation (the analog of a rotation), but instead it is the quantity $x_1^2 + x_2^2 + x_3^2 - x_0^2$ that is Lorentz-invariant. This result is a direct consequence of the second postulate of the theory of relativity that the speed of light in vacuum has a constant value in all inertial frames, coupled with the assumption of homogeneity of space–time* (see Problem 2.7). Mathematically, there are two ways of dealing with this difference. One way is to introduce an imaginary coordinate $x_4 = ict$. Then the above invariance condition is expressed as $x_1^2 + x_2^2 + x_3^2 + x_4^2 =$ constant; this relation is formally a straightforward generalization of the corresponding relation in the 3-dimensional Euclidean space. Thus, the coordinate transformation representing a change of inertial frame (which is the analogue of a rotation) is given in terms of a 4×4 orthogonal matrix $A : X' = AX$; $A^{\mathrm{T}} A = I_{4\times4}$ (obvious notations have been used). There is, however,

*See for e.g. *Introduction to Special Relativity* by Resnik, Wiley 1968

a major difference. This arises from the fact that in the column vectors X and X', the first three elements are real while the fourth one is purely imaginary. This requires that some of the elements of A (viz. A_{i4} and $A_{4i}, i = 1, 2, 3$) are purely imaginary while the rest are real. Thus, unlike the case of 3-dimensional rotations, A is not a real orthogonal matrix but a complex orthogonal matrix. If one keeps this difference in mind, one can deal with the Lorentz transformations and with vectors and tensors in the four-dimensional space–time in a manner that is quite a straightforward generalization of the similar treatment of the 3-dimensional Euclidean space given in the last chapter.

However, there are reasons for not preferring this approach. Firstly, the presence of complex quantities (e.g., keeping track of which ones are complex) is disturbing and unnatural, especially since the physically meaningful quantities must be real. Secondly, the modern literature, wherever it uses Special Theory of Relativity, generally uses real coordinates (x^0, x^1, x^2, x^3), and thus the contact with modern literature is facilitated if one uses real coordinates for space–time. Thirdly, this approach becomes more natural in the General Theory of Relativity where a metric tensor is needed anyway, and which is preferably kept real as physical observables can be formed out of these.

One can stick to real coordinates at the cost of a slight complication, i.e., the introduction of metric tensor $g_{\mu\nu} = \text{Diag}(1, -1, -1, -1)$ (this notation means that $g_{\mu\nu}$ is a diagonal matrix with successive diagonal elements as listed above). Then the Lorentz-invariant $c^2t^2 - x_1^2 - x_2^2 - x_3^2$ can be written as $\sum_\mu \sum_\nu g_{\mu\nu} x^\mu x^\nu$. (The overall sign change for the invariant introduced here is purely conventional). Let under a Lorentz transformation

$$X \equiv \begin{bmatrix} x^0 \\ x^1 \\ x^2 \\ x^3 \end{bmatrix} \qquad \text{transform as}$$

$$X' = \Lambda X \qquad \text{i.e.,} \quad x'^\mu = \Lambda^\mu{}_\alpha x^\alpha, \tag{2.1}$$

Λ being a 4×4 real matrix. The condition satisfied by Λ is obtained in a manner analogous to the 3-dimensional case (see Chapter 1, Section 1.2). The invariant $\sum\sum g_{\mu\nu} x^\mu x^\nu$ can be expressed as $X^\mathrm{T} g X$. Thus, in a Lorentz frame $\sum'$, this quantity is $X'^\mathrm{T} g X'$ and must equal $X^\mathrm{T} g X$ as it is invariant.

$$X'^\mathrm{T} g X' = X^\mathrm{T} g X$$

$$X^\mathrm{T} \Lambda^\mathrm{T} g \Lambda X = X^\mathrm{T} g X \qquad \forall X$$

This requires that*

$$\Lambda^{\mathrm{T}} g \Lambda = g. \tag{2.2}$$

This is the modified form of the condition for Λ, which is the analogue of $A^{\mathrm{T}} A = I$ valid for the case of 3-dimensional rotations.

Taking determinants of both sides one obtains

$$(\det \Lambda)^2 \det g = \det g \neq 0$$

Hence,

$$\det \Lambda = \pm 1 \tag{2.3}$$

The Lorentz transformations with $\det \Lambda = 1$ are called the proper Lorentz transformations; those with $\det \Lambda = -1$ are called the improper Lorentz transformations. The usual Lorentz transformations in which one goes from one inertial frame to another, both having the same handedness and the same time sense, are all special cases of proper Lorentz transformations.

Definition: $g^{\mu\nu}$ is the inverse of the matrix $g_{\mu\nu}$ and equals $g^{\mu\nu} = \mathrm{Diag}(1, -1, -1, -1,) = g_{\mu\nu}$.

Example The Lorentz transformation between two inertial frames $\sum$ and $\sum'$ (with their x and x' axes coincident and the origins coincident at $t = t' = 0$) with $\sum'$ moving with respect to $\sum$ with a velocity $v = \beta c$ along the x-axis is given by

$$\begin{bmatrix} ct' \\ x' \\ y' \\ x' \end{bmatrix} = \begin{bmatrix} \gamma & -\beta\gamma & 0 & 0 \\ -\beta\gamma & \gamma & 0 & 0 \\ 0 & 0 & 1 & 0 \\ 0 & 0 & 0 & 1 \end{bmatrix} \begin{bmatrix} ct \\ x' \\ y' \\ z' \end{bmatrix} \qquad \gamma \equiv \frac{1}{\sqrt{1-\beta^2}}. \tag{2.4}$$

$\det \Lambda$ is obtained by expanding by first two rows and is

$$\det \Lambda = \begin{bmatrix} \gamma & -\beta\gamma \\ -\beta\gamma & \gamma \end{bmatrix} = \gamma^2(1-\beta^2) = +1.$$

An example of a Lorentz transformation with $\det \Lambda = -1$ is given by a similar relation as above except we may replace $z' = z$ in (2.4) by $z' = -z$. This, of course, changes the handedness of the coordinate frame

---

*In this book, we shall often employ the usual set theoretic notations (such as $\in, \exists, \forall, \subset, \cup, \cap$, etc.).

2.2 Covariant and Contravariant 4-Vectors

The procedure for defining a 4-vector is analogous to the one followed in the context of the 3-dimensional Euclidean space. The only difference arises from the fact that now the space is Minkowskian (the metric g has both positive and negative entries) rather than Euclidean. As a result, two kinds of vectors are possible: the covariant and the contravariants ones.

First, we define what is called a contravariant 4-vector. Consider an entity V represented by a quadruplet of numbers $V^\nu \equiv (V^0, V^1, V^2, V^3)$ in an inertial frame $\sum$. Let that entity be represented by (another) quadruplet of numbers $V'^\mu = (V'^0, V'^1, V'^2, V'^3)$ in some other Lorentz frame $\sum'$. Then we say that the entity V is a 4-vector if components of V' and V are related to each other by

$$V'^\mu = \Lambda^\mu{}_\nu V^\nu$$

for every Lorentz frame $\sum'$ related to $\sum$ by a homogeneous Lorentz transformation of Eq. (2.1). (We shall use the usual summation convention that a repeated index occurring as a subscript and a superscript once each is assumed to be summed over.)

In other words, for V to be called a contravariant 4-vector, its Lorentz transformation property must be identical to that of the prototype contravariant 4-vector x^μ.

Example It is shown in standard books on relativity that $p^\mu = (m_0\gamma c, m_0\gamma \mathbf{v})$ with $\gamma = \left(1 - \frac{v^2}{c^2}\right)^{-1/2}$ is a contravariant 4-vector. Note that $p^0 = m_0\gamma c = \frac{m_0\gamma c^2}{c} = \frac{E}{c}$ where E is the relativistic total energy of a free particle. Thus, the total energy and the momentum belong (like time and space) to a single object, i.e., the energy–momentum 4-vector.

2.2.1 Raising and lowering indices

If we look at the Lorentz invariant $c^2t^2 - r^2 = \sum\limits_\mu \sum\limits_\nu x^\mu g_{\mu\nu} x^\nu$, it not only involves the contravariant 4-vector x^μ but also the metric $g_{\mu\nu}$. The above quantity is an analogue of $-\mathbf{x} \cdot \mathbf{x}$ in three dimensions and it should be possible to write it as such. Hence, we define $\sum\limits_\nu g_{\mu\nu} x^\nu \equiv x_\mu$, a new notation, in which the space–time index is shown as a subscript and not as a superscript. We can then write $c^2t^2 - r^2$ as $\sum x^\mu x_\mu \equiv x \cdot x$ and this is the Minkowski analogue of $-\mathbf{x} \cdot \mathbf{x}$ in 3-dimension (please ignore the $-$ve sign, introduced solely by convention). Quantities such as x_μ also occur naturally in Physics and one prefers to deal with them as separate quantities (though they are not independent) rather than their expression in terms of $g_{\mu\nu}$ and x^ν

Definition: In general, given a contravariant 4-vector V^μ, we define

$$V_\mu \equiv \sum_\nu g_{\mu\nu} V^\nu \tag{2.5}$$

and $V_\mu \equiv (V_0, V_1, V_2, V_3) = (V^0, -V^1, -V^2, -V^3)$ are the components of a "covariant 4-vector". Evidently,

$$\begin{aligned} V'_\mu &\equiv \sum_\nu g_{\mu\nu} V^{\nu'} \equiv \sum_\nu g_{\mu\nu} \sum_\alpha \Lambda^\nu{}_\alpha V^\alpha \\ &= \sum g_{\mu\nu} \Lambda^\nu{}_\alpha g^{\alpha\beta} V_\beta \\ &\equiv \sum_\beta \Lambda_\mu{}^\beta V_\beta \end{aligned} \tag{2.6}$$

where $\Lambda_\mu{}^\beta$ has been defined as $\sum_\nu \sum_\alpha g_{\mu\nu} \Lambda^\nu{}_\alpha g^{\alpha\beta}$. Equation (2.6) gives the transformation rule for a covariant 4-vector under a Lorentz transformation.

We note that the new matrix $\Lambda_\mu{}^\beta$ is the inverse transpose of the matrix $\Lambda^\mu{}_\beta$; for

$$\begin{aligned} \Lambda_\mu{}^\beta \Lambda^\mu{}_\gamma &= g_{\mu\nu} \Lambda^\nu{}_\alpha g^{\alpha\beta} \Lambda^\mu{}_\gamma \\ &= (\Lambda^{\mathrm{T}} g \Lambda g^{-1})_\gamma{}^\beta = (g g^{-1})_\gamma{}^\beta = \delta_\gamma{}^\beta. \end{aligned}$$

Also,

$$\begin{aligned} \Lambda^\beta{}_\mu \Lambda_\gamma{}^\mu &= \Lambda^\beta{}_\mu g_{\gamma\eta} \Lambda^\eta{}_\alpha g^{\alpha\mu} = (\Lambda g^{-1} \Lambda^{\mathrm{T}} g)^\beta{}_\gamma \\ &= (\Lambda g \Lambda^{\mathrm{T}} g)^\beta{}_\gamma = (\Lambda g \Lambda^{\mathrm{T}} g \Lambda \Lambda^{-1})^\beta{}_\gamma = (\Lambda g g \Lambda^{-1})^\beta{}_\gamma = \delta^\beta{}_\gamma. \end{aligned}$$

(Here we have used $g^{\gamma\beta} = g^{\beta\gamma}$ and $g^{-1} = g$).

Example We have already seen that $p^\mu = (m_o \gamma c, m_o \gamma \mathbf{v})$ is a contravariant 4-vector. We obtain

$$\begin{aligned} p_\mu = g_{\mu\nu} p^\nu &= p^\mu \quad \text{for} \quad \mu = 0 \\ &= -p^\mu \quad \text{for} \quad \mu = 1, 2, 3. \end{aligned}$$

Hence $p_\mu = (m_o \gamma c, -m_o \gamma \mathbf{v})$ is a covariant 4-vector.

In a similar manner, in electrodynamics, the scalar and the vector potentials combine together as $A^\mu = \left(\frac{\Phi}{c}, \mathbf{A}\right)$ which is a contravariant four vector. Then $A_\mu = \left(\frac{\Phi}{c}, -\mathbf{A}\right)$ is a covariant 4-vector.

Theorem: $\dfrac{\partial}{\partial x^\mu} \equiv \partial_\mu$ transforms as a covariant 4-vector operator.

Proof: Under $x'^\mu = \Lambda^\mu{}_\nu x^\nu$

$$\frac{\partial}{\partial x^\nu} = \frac{\partial x'^\mu}{\partial x^\nu} \frac{\partial}{\partial x'^\mu} = \Lambda^\mu{}_\nu \frac{\partial}{\partial x'^\mu}.$$

We multiply by $\Lambda_\eta{}^\nu$ and sum over ν. We use

$$\Lambda_\eta{}^\nu \Lambda^\mu{}_\nu = \delta^\mu{}_\eta \quad \text{to obtain}$$

$$\Lambda_\eta{}^\nu \frac{\partial}{\partial x^\nu} = \delta^\mu{}_\eta \frac{\partial}{\partial x'^\mu} = \frac{\partial}{\partial x'^\eta},$$

$$\text{i.e.,} \quad \frac{\partial}{\partial x'^\eta} = \Lambda_\eta{}^\nu \frac{\partial}{\partial x^\nu}$$

which is precisely the law for transformation of a covariant 4-vector.

Solved problem 2.1 A photon has energy of 20 MeV in one Lorentz frame $\sum$ and is moving along the positive x-axis in $\sum$. A Lorentz frame $\sum'$ is moving with a velocity 0.6 c along the positive x-axis. Find the energy and the momentum of the photon in $\sum'$.

Solution: $p^\mu = \left[\frac{E}{c}, \mathbf{p}\right]$ forms a contravariant 4-vector; in other words, its transformation properties are identical in form to those of $x^\mu = (ct, \mathbf{x})$. Denoting by p'^μ as the energy momentum 4-vector of the photon in $\sum'$, we can write down the Lorentz transformation as

$$p'^0 = \gamma(p^0 - \beta p_x) \qquad p_y{}' = p_y$$

$$p_x{}' = \gamma(p_x - \beta p^0) \qquad p_z{}' = p_z.$$

Here $\beta = 0.6$, $\gamma = \dfrac{1}{\sqrt{1-(0.6)^2}} = \dfrac{5}{4}$.

$E = 20\,\text{MeV}$; $p_0 = E/c = 20\,\text{MeV}/c$ $\quad |\mathbf{p}| = p_0 = 20\,\text{MeV}/c = p_x$.
Hence,

$$p_0{}' = \frac{E'}{c} = \frac{5}{4}\left[20 - \frac{3}{5} \times 20\right]\text{MeV}/c = 10\,\text{MeV}/c$$

$$p_y{}' = 0,\ p_z' = 0$$

$$p_x{}' = |\mathbf{p}'| = \frac{E'}{c} = 10\,\text{MeV}/c$$

Hence $E' = 10\,\text{MeV}$, $\mathbf{p} = 10\,\text{MeV}/c\,\hat{\mathbf{i}}'$.

2.2.2 Classification of 4-vectors

Unlike the 3-D Euclidean space where the (norm)^2 of a vector $\mathbf{v} \cdot \mathbf{v}$ is always positive, for the Minkowski space 4-vectors $V^\mu V_\mu$ can be positive, negative or zero. As this quantity is Lorentz-invariant, properties implied by the sign of $V^\mu V_\mu$ are maintained in all the Lorentz frames. Hence it is convenient to classify 4-vectors according to the sign of $V^\mu V_\mu$.

Definition:

i) A vector is time-like if $V^\mu V_\mu > 0$
ii) A vector is space-like if $V^\mu V_\mu < 0$
iii) A vector is light-like if $V^\mu V_\mu = 0$
iv) For a vector $V^\mu \equiv (V^0, V^1, V^2, V^3)$; V^0 is called the *time-component* of V^μ and V^1, V^2 and V^3 are called the *space-component* of V^μ; and likewise for $V_\mu = (V_0, V_1, V_2, V_3)$.

We now comment on the origin of these nomenclatures. For the prototype 4-vector $x^\mu = (x^0, x^1, x^2, x^3)$, $x^0 = ct$ is proportional to time and the remaining three components refer to the spatial coordinates. The definitions in (iv) above for the time-component and the space-component originate form this.

The vector $W = (W^0, \mathbf{0})$ has only a nonzero time-component. For this 4-vector $W^\mu W_\mu > 0$. Further, in an another Lorentz frame $W \to W'^\mu = (W'^0, \mathbf{W}')$ with $W'^\mu W'_\mu = W^\mu W_\mu > 0$. Therefore, any 4-vector V^μ with $V^\mu V_\mu > 0$ is called a time-like vector. In fact (see Problem 2.9), in such a case, there exists a Lorentz frame in which $\Lambda V \doteq V' \equiv (V'^0, \mathbf{0})$ has no space components. In any case, for a time like vector V^μ, $|V^0| > |\mathbf{V}|$.

In a similar manner, for a vector V^μ with $V^\mu V_\mu < 0$, there exists a Lorentz frame in which the transformed vector $V'^\mu = (0, \mathbf{V}')$ has only space-components as nonzero (see Problem 2.10). Hence it is called a space-like vector. Note that in this case $|V^0| < |\mathbf{V}|$.

A light quantum, the photon, travelling in free space is characterized by an energy–momentum 4-vector $k^\mu = (k^0, \hbar\mathbf{k})$, where $k^0 = \frac{h\nu}{c}$ and $\mathbf{k}$ is the wave vector. Now, $k^0 = \frac{h\nu}{c} = \frac{h}{\lambda} = \hbar|\mathbf{k}|$. Hence $k^\mu k_\mu = k^{0^2} - |\mathbf{k}|^2\hbar^2 \equiv 0$. Thus the energy–momentum 4-vector of a photon satisfies $k^2 = 0$. Following this, any four-vector V^μ with $V^\mu V_\mu = 0$ is called a light-like vector.

2.3 Lorentz Tensors

We shall now generalize the idea of a 4-vector to that of 4-tensors of an arbitrary rank. This is done in the same manner, as in the previous chapter for the case of 3-D Euclidean space, where we generalized a 3-vector to cartesian tensors of an arbitrary rank. There is one important difference here however. In the 3-D Euclidean space the metric g_{ij} was the identity matrix ($g_{ij} = \delta_{ij} : x^2 = x_i x_j$). Hence, there was no distinction between a covariant and a contravariant vector. In the Minkowski space, $g^{\mu\nu}$ is not a multiple of the identity matrix. Hence by raising/lowering indices as was done in the previous section, we can form covariant and contravariant 4-vectors which are distinct. This distinction persists when we generalize the idea of a 4-vector to that of the 4-tensors. In fact, not only are there covariant and contravariant tensors, now a tensor can also simultaneously have both covariant and contravariant indices.

Definition: A tensor T of rank (n, m) with components $T^{\mu_1\mu_2\cdots\mu_n}_{\nu_1\nu_2\cdots\nu_m}$ is an object such that its components $T'^{\alpha_1\cdots\alpha_n}_{\beta_1\cdots\beta_m}$ in any other Lorentz frame $\sum'$ are related to those in $\sum$ by

$$T'^{\alpha_1\cdots\alpha_n}_{\beta_1\cdots\beta_m} = \Lambda^{\alpha_1}{}_{\mu_1}\Lambda^{\alpha_2}{}_{\mu_2}\cdots\Lambda^{\alpha_n}{}_{\mu_n} \times \Lambda_{\beta_1}{}^{\nu_1}\cdots\Lambda_{\beta_m}{}^{\nu_m} \times T^{\mu_1\cdots\mu_n}_{\nu_1\cdots\nu_m}.$$

Note that this definition is an obvious generalization of the definition of covariant and the contravariant 4-vectors. T is said to have n contravariant and m covariant indices. When $m = 0, n \neq 0$; it is called a contravariant tensor of rank n. When $n = 0, m \neq 0$; it is called a covariant tensor of rank m. When both m and n are nonzero, T is called a mixed tensor of rank (n, m).

Example

(i) $g_{\mu\nu}$ is a covariant tensor of rank two (and is also an invariant tensor). $g^{\mu\nu}$ is a contravariant tensor of rank two. $g^{\mu}_{\nu} = \delta^{\mu}_{\nu}$ is a mixed tensor of rank $(1, 1)$. The proofs of these statements are left as exercises.

(ii) $F_{\mu\nu} = \partial_\mu A_\nu - \partial_\nu A_\mu$ is the field strength covariant tensor of rank two. (The nonvanishing components of $F_{\mu\nu}$ are the electric and magnetic fields. This is so because A_ν is a covariant 4-vector. ∂_μ is a covariant 4-vector operator. Hence $\partial_\mu A_\nu$ is covariant tensor of rank two. So is $-\partial_\nu A_\mu$, and so is their sum. The tensor $g^{\alpha\mu}g^{\beta\nu}F_{\mu\nu} \equiv F^{\alpha\beta}$ is a contravariant tensor of rank two. $g^{\alpha\mu}F_{\mu\nu} = F^{\alpha}{}_{\nu}$ is a mixed tensor of rank $(1, 1)$.

(iii) $\varepsilon^{\mu\nu\lambda\sigma}$ is defined as a generalization of the Levi–Civita symbol in three-dimensions viz., ε_{ijk}. It is totally antisymmetric in all pairs of indices. Further $\varepsilon^{0123} = 1$. (Note : $\varepsilon_{0123} = -1$). It is easy to show, as done for the Levi–Civita tensor in the last chapter, that $\varepsilon^{\mu\nu\lambda\sigma}$ is a pseudotensor of rank four. (Note however that unlike ε_{ijk}, $\varepsilon^{\mu\nu\lambda\sigma}$ is *not* cyclically symmetric. Why?).

2.3.1 Construction of new tensors

Given one or more 4-tensors, one can construct a further set of 4-tensors by processes which are straightforward generalizations of those given for the Cartesian tensors in the last chapter. (We shall list these processes, generally without proofs, as the relevant proofs are straightforward generalizations of those given in the case of the Cartesian tensors.)

(1) Tensor product Given two tensors $T^{(1)}$ and $T^{(2)}$ of ranks (n, m) and (p, q), we can construct a tensor $T^{(3)} = T^{(1)} \otimes T^{(2)}$ of rank $(n+p, m+q)$ whose components are given by

$$T^{(3)}{}^{\alpha_1\alpha_2\cdots\alpha_n\cdots\alpha_{n+p}}_{\beta_1\beta_2\cdots\beta_m\cdots\beta_{m+q}} = T^{(1)\alpha_1\cdots\alpha_n}_{\beta_1\cdots\beta_m} \times T^{(2)}{}^{\alpha_{n+1}\cdots\alpha_{n+p}}_{\beta_{m+1}\cdots\beta_{m+q}}.$$

(2) Contraction Given a mixed tensor of rank (n, m), we equate one contravariant index and any one covariant index and sum over this index from 0 to 3. We thus obtain a tensor of rank $(n-1, m-1)$.

It is important to note that one cannot contract two covariant or two contravariant indices to yield a new tensor, i.e. such an object will not have tensor transformation properties. (See Problem 2.18.)

(3) Differentiation Given a tensor field T of rank (n, m) with components $T^{\alpha_1\cdots\alpha_n}_{\beta_1\cdots\beta_m}(x)$, one can construct a tensor of rank $(n, m+1)$ by taking derivatives, viz.

$$T'^{\,\alpha_1\cdots\alpha_n}_{\beta_1\cdots\beta_m\mu} = \partial_\mu T^{\alpha_1\cdots\alpha_n}_{\beta_1\cdots\beta_m}.$$

This follows from the fact that ∂_μ transforms like a covariant vector differential operator.

(4) Combination of above operations One can, of course, combine above operations, for example:

(a) One can take a tensor product of two tensors and then contract a covariant index in one with a contravariant index in another. e.g. $\varepsilon^{\mu\nu\lambda\sigma}A_\sigma$.

(b) One can take more than one contraction:

$$T^{\alpha\beta\gamma_1\cdots\gamma_n}_{\eta_1\cdots\eta_m\alpha\cdots\eta_p\beta\cdots\eta_q}$$

is a tensor of rank (n, q) obtained from contraction of tensor of rank $(n+2, q+2)$.

(c) One can take derivative and then contract its index to obtain divergence, e.g.,

$$\partial_\mu T^{\alpha\beta} \longrightarrow \partial_\mu T^{\mu\beta}, \quad \text{etc.}$$

Solved problem 2.2 The electromagnetic field strength tensor $F_{\mu\nu}$ is expressed in terms of the 4-vector potential A_μ by $F_{\mu\nu} = \partial_\mu A_\nu - \partial_\nu A_\mu$. Construct $\tilde{F}^{\alpha\beta} = \varepsilon^{\alpha\beta\mu\nu}F_{\mu\nu}$. (i) Show that $\tilde{F}^{\alpha\beta}$ is a contravariant (pseudo) tensor of rank two. (ii) Show that $\tilde{F}$ satisfies $\partial_\alpha \tilde{F}^{\alpha\beta} = 0$. (iii) Show that $F_{\alpha\beta}\tilde{F}^{\alpha\beta}$ can be written as a 4-divergence $\partial_\alpha J^\alpha$ of a quantity J^α, and find J^α.

Solution:

(i) $\varepsilon^{\alpha\beta\mu\nu}$ is a contravariant pseudotensor of rank 4. $F_{\mu\nu}$ is a covariant tensor of rank two. In absence of contraction the product of these would be a mixed tensor of rank (4, 2). Each contraction reduces both ranks by one. Hence rank = (4 − 2, 2 − 2) = (2, 0). Hence $\tilde{F}^{\alpha\beta}$ is a contravariant pseudotensor of rank two.

(ii)
$$\begin{aligned}\partial_\alpha \tilde{F}^{\alpha\beta} &= \partial_\alpha\left[\varepsilon^{\alpha\beta\mu\nu}F_{\mu\nu}\right]\\ &= \varepsilon^{\alpha\beta\mu\nu}\partial_\alpha\left[\partial_\mu A_\nu - \partial_\nu A_\mu\right]\\ &= \varepsilon^{\alpha\beta\mu\nu}\partial_\alpha\partial_\mu A_\nu - \varepsilon^{\alpha\beta\mu\nu}\partial_\alpha\partial_\nu A_\mu.\end{aligned}$$

Consider the first term. $\varepsilon^{\alpha\beta\mu\nu}$ is a tensor antisymmetric in α and μ and $\partial_\alpha\partial_\mu A_\nu$ is a tensor symmetric in α and μ, and α and μ are contracted. Hence the first term vanishes. The second term also vanishes as seen from similar considerations applied to indices α and ν.

(iii) Firstly,

$$\begin{aligned}\tilde{F}^{\alpha\beta} &= \varepsilon^{\alpha\beta\mu\nu} F_{\mu\nu} \\ &= \varepsilon^{\alpha\beta\mu\nu}(\partial_\mu A_\nu - \partial_\nu A_\mu).\end{aligned}$$

But $\varepsilon^{\alpha\beta\mu\nu}\partial_\nu A_\mu$ is obtained by interchanging dummy indices μ and ν; $\varepsilon^{\alpha\beta\mu\nu}\partial_\nu A_\mu = -\varepsilon^{\alpha\beta\mu\nu}\partial_\mu A_\nu$ on account of the antisymmetry of Levi–Civita symbol in (μ, ν). Thus

$$\tilde{F}^{\alpha\beta} = 2\varepsilon^{\alpha\beta\mu\nu}\partial_\mu A_\nu.$$

Now,

$$F_{\alpha\beta}\tilde{F}^{\alpha\beta} = 2(\partial_\alpha A_\beta - \partial_\beta A_\alpha)\varepsilon^{\alpha\beta\mu\nu}\partial_\mu A_\nu.$$

By a reasoning similar to one used earlier, the two terms on the RHS are equal. Hence

$$F_{\alpha\beta}\tilde{F}^{\alpha\beta} = 4\varepsilon^{\alpha\beta\mu\nu}\partial_\alpha A_\beta \partial_\mu A_\nu.$$

Now consider

$$\partial_\alpha\left[\varepsilon^{\alpha\beta\mu\nu}A_\beta\partial_\mu A_\nu\right] = \varepsilon^{\alpha\beta\mu\nu}\partial_\alpha A_\beta\partial_\mu A_\nu + \varepsilon^{\alpha\beta\mu\nu}A_\beta\partial_\alpha\partial_\mu A_\nu.$$

Now $\varepsilon^{\alpha\beta\mu\nu}$ and $\partial_\alpha\partial_\mu A_\nu$ are respectively antisymmetric and symmetric in the two of the contracted indices α and μ. Hence the second term vanishes. This shows that

$$F_{\alpha\beta}\tilde{F}^{\alpha\beta} = \partial_\alpha J^\alpha \quad \text{with} \quad J^\alpha = 4\varepsilon^{\alpha\beta\mu\nu}A_\beta\partial_\mu A_\nu.$$

Problems

On § 2.1

2.1 Show that a Lorentz transformation Λ satisfies

$$\Lambda^{\mathrm{T}} g \Lambda g = 1 = \Lambda^{\mathrm{T}} g^{-1} \Lambda g = \Lambda^{\mathrm{T}} g \Lambda g^{-1}.$$

[Hint: $g = g^{-1}$].

2.2 Given $\Lambda^\mu{}_\nu$ of Eq. (2.4) of § 2.1, obtain $\Lambda_\sigma{}^\tau$. Show by direct calculation that $\Lambda_\sigma{}^\tau\Lambda^\sigma{}_\nu = \delta^\tau{}_\nu$ and $\Lambda^\mu{}_\nu\Lambda_\sigma{}^\nu = \delta^\mu{}_\sigma$.

2.3 Expressing $\Lambda_\sigma{}^\tau \equiv \Lambda'^\tau{}_\sigma$, show that $\Lambda' = g\Lambda^T g$. Hence show that $\Lambda_\sigma{}^\tau \Lambda^\sigma{}_\nu = \delta^\tau{}_\nu$ now reads $\Lambda'\Lambda = 1$. Prove this using the earlier expression for Λ'.

2.4 Construct the Lorentz transformation Λ from a Lorentz frame $\sum$ to a Lorentz frame $\sum'$ that is moving with speed $\beta\left[\frac{\hat{\mathbf{i}}+\hat{\mathbf{j}}}{\sqrt{2}}\right]$ with respect to $\sum$. Do so by using a rotation of axes followed by a Lorentz transformation.

2.5 Formulate the matrices for pure rotations (R), parity transformation (P), time-reversal transformation (T) as special case of $\Lambda^\mu{}_\nu$.

2.6 Show that an improper Lorentz transformation with $\det\Lambda = -1$ can be written either as $\Lambda = P\Lambda_{(P)}$ or $\Lambda = T\Lambda_{(P)}$ where P and T are matrices constructed in Problem 2.5 and $\Lambda_{(P)}$ is a proper time sense-preserving (orthochronus) Lorentz transformation with $\det\Lambda_P = +1$.

2.7 Using homogenity of space–time, isotropy of space and constancy of speed of light in vacuum, argue that $x_\mu x^\mu = x'_\mu x'^\mu$ where x^μ and x'^μ refer to space–time coordinates of an event in two Lorentz frames $\sum$ and $\sum'$.

On § 2.2

2.8 Show that $V^\mu V_\mu$ is a Lorentz-invariant (scalar).

2.9 Show that for a time-like vector $V^\mu(|V^0| > |\mathbf{V}|)$, it is always possible to find a Lorentz frame $\sum'$ such that $V'^\mu = (V'^0, \mathbf{0})$, i.e. space-like components of V' vanish. Is such a frame unique? What does the result imply for the energy–momentum 4-vector p^μ of a massive particle?
$(p^\mu p_\mu = m^2 > 0)$.

2.10 Show that for a space-like vector $W^\mu(|W^0| < |\mathbf{W}|)$, it is always possible to find a Lorentz frame $\sum'$ such that $W'^\mu = (0, \mathbf{W}')$, i.e., the time components of W' vanishes.

2.11 Consider the electromagnetic vector potential 4-vector $A^\mu = \left[\frac{\Phi}{c}, \mathbf{A}\right]$. Classify whether A^μ is space-like/time-like for following circumstances:

i) A^μ for a stationary point charge in Coulomb gauge.
ii) A^μ for a current-carrying wire (I = constant) in Coulomb gauge.
iii) A^μ for a plane wave in radiation gauge.

Does the space-like/time-like/light-like nature of $A^\mu(x)$ in these cases depend on gauge? on x^μ?

2.12 Let $p_1, p_2, \cdots, p_n$ be the energy–momentum 4-vectors of particles each of mass m.

Show that $\left[\sum_{i=1}^{n} p_i\right]^2 \geq (nm)^2$.

[Hint: Show that $\sum_{i=1}^{n} p_i$ is a time-like vector. Use the result of Problem 2.9].

2.13 Verify explicitly that the light-like vector $(k, 0, 0, k)$ goes into another light-like vector under the Lorentz transformation Λ of Eq. (2.4) of § 2.1.

2.14 For a vector k find all linearly independent solutions to $k^\mu \varepsilon_\mu(k) = 0$ when k is (i) light-like (ii) time-like.

On § 2.3

2.15 Show that $g_{\mu\nu}$, $g^{\mu\nu}$ and $\delta^\mu{}_\nu$ are the three invariant tensors of rank (0, 2); (2, 0); (1, 1); respectively.

2.16 Show that $\varepsilon^{\mu\nu\lambda\sigma} = -\varepsilon_{\mu\nu\lambda\sigma} = -\varepsilon^{\nu\lambda\sigma\mu}$.

2.17 Show that $\varepsilon^{\mu\nu\lambda\sigma}$ is an invariant pseudotensor of rank (4, 0).

2.18 Let $T_{\mu\nu}$ be a covariant tensor of rank 2. Show then that $\sum_{\mu=0}^{3} T_{\mu\mu}$ is not a Lorentz scalar in general. Verify this with the explicit example of $T_{\mu\nu} \equiv x_\mu x_\nu$. Generalize the result.

2.19 Given the equation $k^\mu G_{\mu\nu}(k) = 0$ for all k, show that $G_{\mu\nu}(k)$ must be expressible in the form $G_{\mu\nu}(k) = (g_{\mu\nu}k^2 - k_\mu k_\nu)H(k^2)$, $(k^2 \neq 0)$.

3 Curvilinear Coordinates

While the cartesian coordinates are, by far, the simplest ones to use generally, there are many problems in Physics, where other sets of coordinate systems are more useful. This happens, typically, when the physical situation has a certain symmetry (such as spherical symmetry, cylindrical symmetry, etc.) For example, consider the problem of obtaining electric field of a uniformly charged spherical shell by directly solving Poisson equation, $\nabla^2\phi = -\rho/\varepsilon_o$. (For such a simple example, there are other powerful methods but we are considering this case as an illustrative example.) In Cartesian coordinates one would, then, have to solve the partial differential equation in 3 variables

$$\left[\frac{\partial^2}{\partial x^2} + \frac{\partial^2}{\partial y^2} + \frac{\partial^2}{\partial z^2}\right] \phi(x, y, z) = -\rho/\varepsilon_o$$

subject to the boundary conditions (which has a certain symmetry.) On the other hand, one could notice at the outset that the system has a spherical symmetry which is conveniently exploited by taking the center of the sphere as origin. Then $\phi(x, y, z)$ must be function only of r, the radial distance, i.e., a function of a single variable and this simplyfies the problem. But for this symmetry to be fully exploited, one needs to express ∇^2 in spherical polar coordinates. One of the main purposes of this chapter is to express differential operators in curvilinear coordinates, so useful in a large number of problems in Physics. We shall, however, confine the discussion only to orthogonal curvilinear coordinates in this book.

Before proceeding further, we give illustrations of some of the common problems in Physics which use curvilinear coordinates. To obtain solutions for planetary orbits, one uses spherical polar coordinates. To obtain the solution for energy levels and wavefunctions of a hydrogen atom in Quantum Mechanics, one also uses spherical polar coordinates. Solution for electromagnetic fields in coaxial cables is facilitated by the use of cylindrical coordinates. The hydrogen atom problem in Quantum Mechanics, mentioned above, can also be done using what are know as parabolic coordinates.

3.1 Coordinate Curves

First let us look at the familiar cartesian coordinate system from a somewhat different angle because the features we are going to note down now remain valid for any orthogonal coordinate system. Normally, one looks at the cartesian coordinates as being given by a set of three mutually perpendicular lines called axes. The directions of the lines are, in a given cartesian coordinate frame, fixed once and for all; the directions of unit vectors do not change from point to point. In other words, through a point (a, b, c) there are three mutually perpendicular lines passing: (i) $z = c$, $y = b$; (ii) $z = c$, $x = a$; (iii) $x = a$, $y = b$. These three lines are parallel to the three coordinate axes at any point. Unit tangents to these lines give the unit vectors and the triad of unit vectors at one point is parallel to the triad of unit vectors at another point. Unit tangents at P are given by ∇x, ∇y, ∇z at P and are along $\hat{\mathbf{i}}, \hat{\mathbf{j}}, \hat{\mathbf{k}}$.

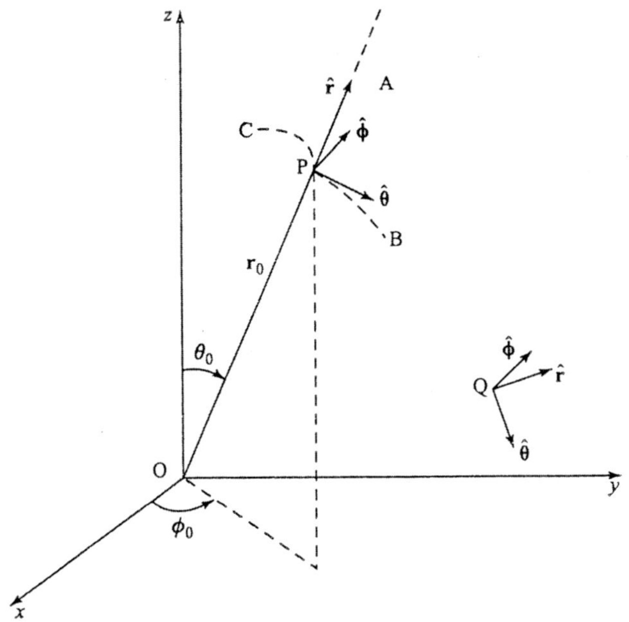

Fig. 3.1 *Spherical polar coordinates: coordinate curves and the triad of unit vectors.*

When one makes transition to an orthogonal curvilinear coordinate system, there still continue to be three coordinate curves passing through a given point P. The unit tangent at P to the three curves do form an orthonormal triad (for an orthogonal curvilinear system.) But the coordinate curves need no longer be straight lines in general, so that the direction of the unit vectors of the triad can alter continuously as the point P is altered. An example is given below of the spherical polar coordinate system. Consider the point P given by spherical polar coordinates $r = r_0$, $\theta = \theta_0$, $\phi = \phi_0$. The three coordinate curves are:

(i) PA: $\theta = \theta_0$, $\phi = \phi_0$
(ii) PB: $r = r_0$, $\phi = \phi_0$
(iii) PC: $\theta = \theta_0$, $r = r_0$

Along each coordinate curve only one coordinate is changing: r along PA, θ along PB and ϕ along PC. The triad of orthonormal units tangents is also shown (arrows point in the direction of increasing r, θ and φ respectively.) and these are denoted by $\hat{\mathbf{r}}, \hat{\boldsymbol{\theta}}$ and $\hat{\boldsymbol{\varphi}}$. The triad of orthonormal unit vectors is shown also at another point Q and it may be noticed that the respective unit vectors are not parallel to those at P.

In the case of a general orthogonal coordinate system (u, v, w) the situation is as shown in Fig. 3.2. Through any point P pass three coordinate curves: along one only u increases/decreases, v and w remain constant; along another v increases/decreases and w and u remain constant, and so on. We shall call these the u-coordinate line, v-coordinate line and the w-coordinate line through P respectively. Unit vectors along the three coordinate lines are tangents to the three coordinate lines along the sense in which the coordinate increase. As we shall see, these are given by $\frac{\nabla u}{|\nabla u|}, \frac{\nabla v}{|\nabla v|}, \frac{\nabla w}{|\nabla w|}$ respectively. To see this consider v and w coordinate lines through any point P. Along these lines u is constant. Hence ∇u is perpendicular to the unit tangent vectors $\hat{\mathbf{v}}$ and $\hat{\mathbf{w}}$ and v and w coordinate lines. But, by assumption, this is an orthogonal coordinate system and hence $\hat{\mathbf{u}}$ is also perpendicular to $\hat{\mathbf{v}}$ and $\hat{\mathbf{w}}$. Moreover, $\hat{\mathbf{u}}$ points along the direction in which u is increasing. Thus, $\hat{\mathbf{u}}$ and $\frac{\nabla u}{|\nabla u|}$ have identical properties (mentioned above) that fix $\hat{\mathbf{u}} = \frac{\nabla u}{|\nabla u|}$. In a similar manner, we also have $\hat{\mathbf{v}} = \frac{\nabla u}{|\nabla v|}, \hat{\mathbf{w}} = \frac{\nabla u}{|\nabla w|}$.

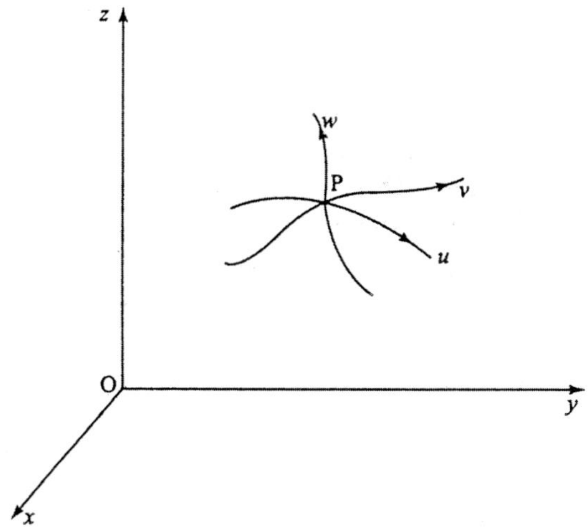

Fig. 3.2 *Coordinate curves through a point P.*

3.2 Scale Factors

Consider first the cartesian Coordinate system. If one considers two neighbouring points on the x-axis $(x, 0, 0)$ and $(x + dx, 0, 0)$, the x-coordinate difference dx and the distance between these two points ds are identical, i.e., $\frac{\partial x}{\partial s} = 1$ along the x-coordinate line. This property that the coordinate difference between neighbouring points along a given coordinate line equals the distance between them may not generally hold for an arbitrary curvilinear coordinate system. For example, if one considers two neighboring points P and Q given in spherical polar coordinates by (r, θ, ϕ) and $(r, \theta + d\theta, \phi)$, the coordinate separation $d\theta$ and the distance ds between them are related by $d\theta = ds/r$, or,

$$\left.\frac{\partial\theta}{\partial s}\right|_{\substack{r=\text{const}\\ \varphi=\text{const}}} = \frac{1}{r} \neq 1 \quad \text{generally.}$$

We, therefore, need to define a scale factor for each coordinate line that gives the rate at which distance s changes per unit coordinate difference.

In other words, we shall define the scale factors by h_u, h_v, h_w by

$$\begin{aligned} h_u^{-1} &= \left.\frac{\partial u}{\partial s}\right|_{\substack{v=\text{const}\\ w=\text{const}}} \\ h_v^{-1} &= \left.\frac{\partial v}{\partial s}\right|_{\substack{u=\text{const}\\ w=\text{const}}} \\ h_w^{-1} &= \left.\frac{\partial w}{\partial s}\right|_{\substack{u=\text{const}\\ v=\text{const}}} \end{aligned} \tag{3.1}$$

The scale factors may, generally, be functions of coordinates u, v, w themselves, i.e., they may change along respective coordinate line or may be different for different coordinate lines corresponding to a given coordinates. Thus, in the above example, $h_\theta = r$ and changes from one r-coordinate surface (sphere) to another. It is easy to see that $h_r = 1, h_\phi = r\sin\theta$. (A scale factor may, in special cases, be just unity as in the case of h_r. But not all the three scale factors are equal to unity everywhere except for the cartesian coordinate system. (see Problem 3.12).

We now derive expressions for scale factors. Following the logic of Problem 3.21,

$$\left.\frac{\partial u}{\partial s}\right|_{\substack{v=\text{const}\\ w=\text{const}}} = \nabla u.\hat{\mathbf{u}} = |\nabla u|\hat{\mathbf{u}}.\hat{\mathbf{u}} = |\nabla u|$$

where we have used $\hat{\mathbf{u}} = \nabla u/|\nabla u|$.

Thus, we have

$$h_u^{-1} = |\nabla u| = \left\{\left[\frac{\partial u}{\partial x}\right]^2 + \left[\frac{\partial u}{\partial y}\right]^2 + \left[\frac{\partial u}{\partial z}\right]^2\right\}^{1/2} \tag{3.2}$$

and similar expressions for h_v and h_w. These expressions are useful to compute scale factors when u, v, w are given as functions of x, y, z.

We shall now derive alternate expressions for the scale factors useful in computing them when x, y, z are given as functions of u, v, w.

Consider again the vector

$$\left.\frac{\partial \mathbf{r}}{\partial u}\right|_{v,w} = \frac{\partial x}{\partial u}\hat{\mathbf{i}} + \frac{\partial y}{\partial u}\hat{\mathbf{j}} + \frac{\partial z}{\partial u}\hat{\mathbf{k}}.$$

Now modulus of this vector $|\partial \mathbf{r}/\partial u|$ can be obtained by considering the neighbouring points on the u-coordinate line $P = (u, v, w)$ and $Q(u + du, v, w)$ with $\mathbf{PQ} = d\mathbf{r}$ and $|PQ| = |d\mathbf{r}| = ds$.

Hence

$$\left|\frac{\partial \mathbf{r}}{\partial u}\right|_{v,w} = \frac{|d\mathbf{r}|}{du} = \left.\frac{\partial s}{\partial u}\right|_{v,w} = h_u; \qquad du > 0$$

Thus, by definition, we have,

$$h_u = \left|\frac{\partial \mathbf{r}}{\partial u}\right| = \left\{\left[\frac{\partial x}{\partial u}\right]^2 + \left[\frac{\partial y}{\partial u}\right]^2 + \left[\frac{\partial z}{\partial u}\right]^2\right\}^{1/2} \tag{3.3}$$

and similar expressions for h_v and h_w.

Example 1 Calculate the scale factors for spherical polar coordinates.

Solution: We have

$$x = r\sin\theta\cos\phi \qquad y = r\sin\theta\sin\phi \qquad z = r\cos\theta.$$

Expressions of (3.3) are useful here. We obtain

$$\begin{aligned}
h_r^2 &= \left[\frac{\partial x}{\partial r}\right]^2 + \left[\frac{\partial y}{\partial r}\right]^2 + \left[\frac{\partial z}{\partial r}\right]^2 \\
&= \sin^2\theta\cos^2\phi + \sin^2\theta\sin^2\phi + \cos^2\theta \\
&= 1 \\
h_\theta^2 &= \left[\frac{\partial x}{\partial \theta}\right]^2 + \left[\frac{\partial y}{\partial \theta}\right]^2 + \left[\frac{\partial z}{\partial \theta}\right]^2 \\
&= r^2\cos^2\theta\ \cos^2\phi + r^2\cos^2\theta\sin^2\phi + r^2\sin^2\theta \\
&= r^2 \\
h_\phi^2 &= \left[\frac{\partial x}{\partial \phi}\right]^2 + \left[\frac{\partial y}{\partial \phi}\right]^2 + \left[\frac{\partial z}{\partial \phi}\right]^2 \\
&= r^2\sin^2\theta\sin^2\phi + r^2\sin^2\theta\cos^2\phi \\
&= r^2\sin^2\theta.
\end{aligned}$$

As scale factors are positive, $h_r = 1, h_\theta = r, h_\phi = r\sin\theta$. We may alternatively calculate these using (3.2).

We have

$$r = \sqrt{x^2 + y^2 + z^2}; \qquad \theta = \tan^{-1}\frac{\sqrt{x^2+y^2}}{z}; \qquad \phi = \tan^{-1}\frac{y}{x}.$$

$$\frac{1}{h_r^2} = \left[\frac{\partial r}{\partial x}\right]^2 + \left[\frac{\partial r}{\partial y}\right]^2 + \left[\frac{\partial r}{\partial z}\right]^2 = \left[\frac{x}{r}\right]^2 + \left[\frac{y}{r}\right]^2 + \left[\frac{z}{r}\right]^2 = 1;$$

$$\frac{1}{h_\theta^2} = \left[\frac{\partial \theta}{\partial x}\right]^2 + \left[\frac{\partial \theta}{\partial y}\right]^2 + \left[\frac{\partial \theta}{\partial z}\right]^2 = \frac{1}{r^2}; \qquad \text{(Prove !)}$$

$$\frac{1}{h_\phi^2} = \left[\frac{\partial \phi}{\partial x}\right]^2 + \left[\frac{\partial \phi}{\partial y}\right]^2 + \left[\frac{\partial \phi}{\partial z}\right]^2 = \left[\frac{-y/x^2}{1+\dfrac{y^2}{x^2}}\right]^2 + \left[\frac{1/x}{1+\dfrac{y^2}{x^2}}\right]^2$$

$$= \frac{1}{x^2+y^2} = \frac{1}{r^2\sin^2\theta};$$

yielding the same results.

Example 2 The parabolic coordinate system (useful in solving Schroedinger equation for the H-atom and Coulomb scattering problems) is defined by

$$x = uvw \qquad y = uv\sqrt{1-w^2} \qquad z = \frac{1}{2}(u^2 - v^2) \qquad |w| \leq 1.$$

(i) Solve for u, v, w in terms of x, y, z.

(ii) Obtain scale factors h_u, h_v, h_w (as function of u, v, w).

(iii) Obtain unit vectors along coordinate lines at (u, v, w).

Solution:

(i) $x^2 + y^2 = u^2 v^2$.

$$(x^2+y^2) + z^2 = r^2 = u^2v^2 + \frac{1}{4}(u^4 + v^4 - 2u^2v^2) = \frac{1}{4}(u^2+v^2)^2$$

$$r = \frac{1}{2}(u^2 + v^2).$$

$$u^2 = r + z; \quad v^2 = r - z; \quad w = \frac{x}{uv} = \frac{x}{\sqrt{x^2+y^2}}.$$

(ii) Here x, y, z are more easily expressed in terms of u, v, w and we shall use these expressions in (3.3) to obtain h_u, h_v, h_w.

$$h_u = \left\{\left[\frac{\partial x}{\partial u}\right]^2 + \left[\frac{\partial y}{\partial u}\right]^2 + \left[\frac{\partial z}{\partial u}\right]^2\right\}^{1/2} = \left\{v^2 w^2 + v^2(1-w^2) + u^2\right\}^{1/2}$$

$$= \left\{u^2 + v^2\right\}^{1/2}.$$

$$h_v = \left\{\left[\frac{\partial x}{\partial v}\right]^2 + \left[\frac{\partial y}{\partial v}\right]^2 + \left[\frac{\partial z}{\partial v}\right]^2\right\}^{1/2} = \left\{(uw)^2 + u^2(1-w^2) + v^2\right\}^{1/2}$$

$$= \left\{u^2 + v^2\right\}^{1/2}$$

$$h_w = \left\{\left[\frac{\partial x}{\partial w}\right]^2 + \left[\frac{\partial y}{\partial w}\right]^2 + \left[\frac{\partial z}{\partial w}\right]^2\right\}^{1/2} = \left\{u^2 v^2 + \frac{u^2 v^2 w^2}{(1-w^2)}\right\}^{1/2}$$

$$= \frac{uv}{(1-w^2)^{1/2}}.$$

(iii) The unit vectors may be calculated by using

$$\hat{\mathbf{u}} = \frac{\nabla u}{|\nabla u|} = h_u \nabla u\,; \quad \hat{\mathbf{v}} = h_v \nabla v\,; \quad \hat{\mathbf{w}} = h_w \nabla w.$$

Alternatively one notes that the vector

$\left.\dfrac{\partial \mathbf{r}}{\partial u}\right|_{v,w}$ is along the direction of $\hat{\mathbf{u}}$ and has magnitude h_u.

Hence

$$\hat{\mathbf{u}} = \frac{1}{h_u}\frac{\partial \mathbf{r}}{\partial u} = \frac{1}{(u^2+v^2)^{1/2}}\left[\frac{\partial x}{\partial u}\hat{\mathbf{i}} + \frac{\partial y}{\partial u}\hat{\mathbf{j}} + \frac{\partial z}{\partial u}\hat{\mathbf{k}}\right]$$

$$= \frac{vw\,\hat{\mathbf{i}} + v\sqrt{1-w^2}\,\hat{\mathbf{j}} + u\,\hat{\mathbf{k}}}{(u^2+v^2)^{1/2}}$$

$$= \frac{uvw\,\hat{\mathbf{i}} + uv\sqrt{1-w^2}\,\hat{\mathbf{j}} + u^2\,\hat{\mathbf{k}}}{u(u^2+v^2)^{1/2}}$$

$$= \frac{x\,\hat{\mathbf{i}} + y\,\hat{\mathbf{j}} + (r+z)\,\hat{\mathbf{k}}}{\sqrt{r+z}(2r)^{1/2}}.$$

In a similar manner,

$$\hat{\mathbf{v}} = \frac{1}{\sqrt{r-z}(2r)^{1/2}}\left\{x\,\hat{\mathbf{i}} + y\,\hat{\mathbf{j}} + (z-r)\,\hat{\mathbf{k}}\right\}.$$

$$\hat{\mathbf{w}} = \frac{y}{\sqrt{x^2+y^2}}\hat{\mathbf{i}} - \frac{x}{\sqrt{x^2+y^2}}\hat{\mathbf{j}}.$$

3.3 Gradient, Divergence and Curl in Curvilinear Coordinates

3.3.1 Gradient

Consider a function $\Phi(u, v, w)$. We want to define the gradient of Φ in terms of the unit vectors $\hat{\mathbf{u}}$, $\hat{\mathbf{v}}$, $\hat{\mathbf{w}}$ and partial derivatives of Φ with respect to u, v, w. Now we know that $\nabla\Phi \cdot \hat{\mathbf{n}}$ gives the directional derivative $\frac{\partial\Phi}{\partial s}$ along the unit vector $\hat{\mathbf{n}}$. Setting $\hat{\mathbf{n}} = \hat{\mathbf{u}}$,

$$\nabla\Phi \cdot \hat{\mathbf{u}} = \frac{d\Phi}{ds} \quad \text{along } u\text{-coordinate line.}$$

Now, along u-coordinate line $ds = h_u du$. Hence, along u-coordinate line

$$\frac{d\Phi}{ds} = \frac{1}{h_u} \left.\frac{\partial\Phi}{\partial u}\right|_{v,w} = \nabla\Phi \cdot \hat{\mathbf{u}}.$$

In a similar manner we obtain components of $\nabla\Phi$ along $\hat{\mathbf{v}}$ and $\hat{\mathbf{w}}$ and assemble $\nabla\Phi$:

$$\begin{aligned}\nabla\Phi &= \frac{1}{h_u}\frac{\partial\Phi}{\partial u}\hat{\mathbf{u}} + \frac{1}{h_v}\frac{\partial\Phi}{\partial v}\hat{\mathbf{v}} + \frac{1}{h_w}\frac{\partial\Phi}{\partial w}\hat{\mathbf{w}} \\ &\equiv \left[\hat{\mathbf{u}}\frac{1}{h_u}\frac{\partial}{\partial u} + \hat{\mathbf{v}}\frac{1}{h_v}\frac{\partial}{\partial v} + \hat{\mathbf{w}}\frac{1}{h_w}\frac{\partial}{\partial w}\right]\Phi(u, v, w).\end{aligned}$$

Or relabelling the curvilinear coordinates as u_1, u_2, u_3, we can write ∇ in a compact notation

$$\nabla = \sum_{i=1}^{3} \hat{\mathbf{u}}_i \frac{1}{h_i}\frac{\partial}{\partial u_i}. \tag{3.4}$$

$\nabla\phi$ at P is perpendicular to the surface of $\phi =$ constant through P. Its magnitude gives the rate of change of φ in this perpendicular direction (see Problem 3.21).

3.3.2 Divergence

Consider a vector field $\mathbf{B}$ expressed in terms of the unit vectors of an orthogonal curvilinear coordinate system:

$$\mathbf{B} = \sum_{i=1}^{3} B_i(u_1, u_2, u_3)\,\hat{\mathbf{u}}_i \equiv \sum_{i=1}^{3} B_i(u)\,\hat{\mathbf{u}}_i. \tag{3.5}$$

Then

$$\nabla \cdot \mathbf{B} = \sum_{i=1}^{3} \nabla \cdot (B_i\hat{\mathbf{u}}_i). \tag{3.6}$$

It suffices to evaluate divergence of a vector $B_i\hat{\mathbf{u}}_j$ parallel to *one* of the unit vectors $\hat{\mathbf{u}}_j$ at a time.

We shall evaluate, say, $\nabla \cdot [B_1(u)\hat{\mathbf{u}}_1]$ by applying the Gauss' divergence theorem

$$\int_{dV} \nabla \cdot (B_1 \hat{\mathbf{u}}_1) dV = \oint B_1 \hat{\mathbf{u}}_1 \cdot \hat{\mathbf{n}} da$$

to an infinitesimal volume element conveniently chosen, in the shape of (approximately) a rectangular parellelopiped with sides parallel to the coordinate lines. See Fig. 3.3*a*.

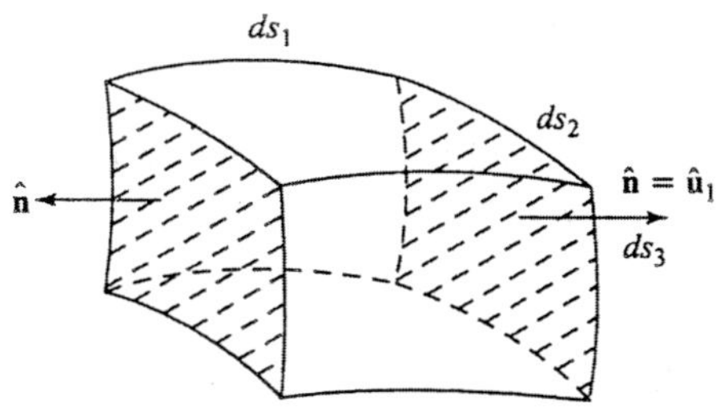

Fig. 3.3*a* *Infinitesimal volume element for the application of Gauss' theorem*

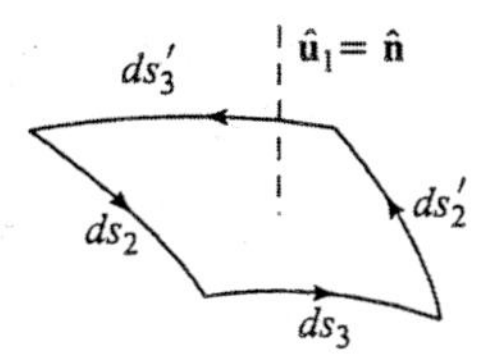

Fig. 3.3*b* *Infinitesimal area element for the application of Stokes' theorem*

Now, of the six surfaces, $\hat{\mathbf{u}}_1 \cdot \hat{\mathbf{n}}$ is nonvanishing (± 1) only for the two shaded ones. We note that the surface elements for these are $da = ds_2 ds_3 = h_2(u)h_3(u)du_2du_3$ and are themselves generally different for these two surfaces as the scale factors are generally u-dependent.

In view of the fact that the volume element is infinitesimal, we regard $\nabla.(B_1\hat{\mathbf{u}}_1)$ constant over it. We also regard B_1 to have uniform values over each of the shaded surfaces. As $B_1(u)$ may depend on u_1, its values for the two shaded surfaces could however be different. (To put it more clearly, the variation of $\nabla \cdot \mathbf{B}$ in the volume and variation of B_1 over each of the shaded surfaces leads to corrections that vanish as the ds_1, ds_2 and ds_3 tend to zero, but the variation of B_1 along $\hat{\mathbf{u}}_1$ contributes). Hence

$$\nabla \cdot (B_1\hat{\mathbf{u}}_1)dV = \big[h_2(u_1 + du_1, u_2, u_3)h_3(u_1 + du_1, u_2, u_3)B_1(u_1 + du_1, u_2, u_3) - h_2(u_1, u_2, u_3)h_3(u_1, u_2, u_3)B_1(u_1, u_2, u_3)\big]du_2du_3,$$

$$\nabla \cdot (B_1\hat{\mathbf{u}}_1)h_1h_2h_3du_1du_2du_3 = \frac{\partial}{\partial u_1}(h_2h_3B_1)du_1du_2du_3.$$

Hence

$$\nabla \cdot (B_1\hat{\mathbf{u}}_1) = \frac{1}{h_1h_2h_3} \frac{\partial}{\partial u_1}(h_2h_3B_1).$$

One may repeat the same procedure for other terms in (3.6) so that

$$\nabla \cdot \mathbf{B} = \frac{1}{h_1 h_2 h_3} \sum_{i=1}^{3} \frac{\partial}{\partial u_i} \left[\frac{h_1 h_2 h_3 B_i}{h_i} \right]. \tag{3.7}$$

3.3.3 Curl

We shall now obtain the curl of **B** of Eq. (3.5) using the Stokes' theorem

$$\int_S \nabla \times \mathbf{B} \cdot \hat{\mathbf{n}}\, da = \oint \mathbf{B} \cdot \vec{\mathbf{dl}} \tag{3.8}$$

We shall choose S to be an infinitesimal rectangular surface with normal chosen along $\hat{\mathbf{u}}_1$ and sides parallel to $\hat{\mathbf{u}}_2$ and $\hat{\mathbf{u}}_3$ (see Fig. 3.3*b*). The area is $da = h_2 h_3 du_2 du_3$.

Then,

$$\begin{aligned}
(\nabla \times \mathbf{B})_1 h_2 h_3 du_2 du_3 &= B_2(u_1, u_2, u_3) h_2(u_1, u_2, u_3) du_2 \\
&\quad + B_3(u_1, u_2 + du_2, u_3) h_3(u_1, u_2 + du_2, u_3) du_3 \\
&\quad - B_2(u_1, u_2 + du_2, u_3 + du_3) h_2(u_1, u_2 + du_2, u_3 + du_3) du_2 \\
&\quad - B_3(u_1, u_2, u_3 + du_3) h_3(u_1, u_2, u_3 + du_3) du_3 \\
&= -du_2 du_3 \frac{\partial}{\partial u_3}(h_2 B_2) + du_3 du_2 \frac{\partial}{\partial u_2}(h_3 B_3) \\
&\quad + \text{ terms that vanish faster as } du_k \to 0.
\end{aligned}$$

(We have ignored the variation of B_i along a given side for the same reasons as in the previous subsection.)

Thus,

$$(\nabla \times \mathbf{B})_1 = -\frac{1}{h_2} \frac{1}{h_3} \frac{\partial}{\partial u_3}(h_2 B_2) + \frac{1}{h_2 h_3} \frac{\partial}{\partial u_2}(h_3 B_3).$$

One can obtain the other components of $\nabla \times \mathbf{B}$ along similar lines. The net result is then

$$\nabla \times \mathbf{B} = \frac{1}{h_1 h_2 h_3} \begin{bmatrix} h_1 \hat{\mathbf{u}}_1 & h_2 \hat{\mathbf{u}}_2 & h_3 \hat{\mathbf{u}}_3 \\ \dfrac{\partial}{\partial u_1} & \dfrac{\partial}{\partial u_2} & \dfrac{\partial}{\partial u_3} \\ h_1 B_1 & h_2 B_2 & h_3 B_3 \end{bmatrix}.$$

Example Gradient, divergence and curl and the Laplacian in spherical polar coordinates:

(i) $$\nabla = \sum \hat{\mathbf{u}}_i \frac{1}{h_i} \frac{\partial}{\partial u_i} = \frac{\hat{\mathbf{r}}}{h_r} \frac{\partial}{\partial r} + \frac{\hat{\boldsymbol{\theta}}}{h_\theta} \frac{\partial}{\partial \theta} + \frac{\hat{\boldsymbol{\phi}}}{h_\phi} \frac{\partial}{\partial \phi}$$

$$= \hat{\mathbf{r}}\frac{\partial}{\partial r} + \hat{\boldsymbol{\theta}}\frac{1}{r}\frac{\partial}{\partial \theta} + \hat{\boldsymbol{\phi}}\frac{1}{r\sin\theta}\frac{\partial}{\partial \phi}.$$

(ii) Let $\mathbf{A} = A_r\hat{\mathbf{r}} + A_\theta\hat{\boldsymbol{\theta}} + A_\phi\hat{\boldsymbol{\phi}} = A_1\hat{\mathbf{u}}_1 + A_2\hat{\mathbf{u}}_2 + A_3\hat{\mathbf{u}}_3$

$$\nabla \cdot \mathbf{A} = \frac{1}{h_r h_\theta h_\phi}\left[\frac{\partial}{\partial r}(h_\theta h_\phi A_r) + \frac{\partial}{\partial \theta}(h_\phi h_r A_\theta) + \frac{\partial}{\partial \phi}(h_r h_\theta A_\phi)\right]$$

$$= \frac{1}{r^2\sin\theta}\left[\frac{\partial}{\partial r}(r^2\sin\theta\, A_r) + \frac{\partial}{\partial \theta}(r\sin\theta\, A_\theta) + \frac{\partial}{\partial \phi}(r A_\phi)\right]$$

$$= \frac{\partial A_r}{\partial r} + \frac{1}{r}\frac{\partial A_\theta}{\partial \theta} + \frac{1}{r\sin\theta}\frac{\partial A_\phi}{\partial \phi} + \frac{2}{r}A_r + \frac{\cot\theta}{r}A_\theta.$$

(iii) $$\nabla \times \mathbf{A} = \frac{1}{h_r h_\theta h_\phi}\begin{vmatrix} h_r\hat{\mathbf{r}} & h_\theta\hat{\boldsymbol{\theta}} & h_\phi\hat{\boldsymbol{\phi}} \\ \dfrac{\partial}{\partial r} & \dfrac{\partial}{\partial \theta} & \dfrac{\partial}{\partial \phi} \\ h_r A_r & h_\theta A_\theta & h_\phi A_\phi \end{vmatrix}$$

$$= \frac{1}{r^2\sin\theta}\begin{vmatrix} \hat{\mathbf{r}} & r\hat{\boldsymbol{\theta}} & r\sin\theta\,\hat{\boldsymbol{\phi}} \\ \dfrac{\partial}{\partial r} & \dfrac{\partial}{\partial \theta} & \dfrac{\partial}{\partial \phi} \\ A_r & rA_\theta & r\sin\theta\, A_\phi \end{vmatrix}$$

Problems

On § 3.1

3.1 Obtain relations (both ways) between $\hat{\mathbf{i}}, \hat{\mathbf{j}}, \hat{\mathbf{k}}$ and $\hat{\mathbf{r}}, \hat{\boldsymbol{\theta}}, \hat{\boldsymbol{\phi}}$ (of the spherical polar coordinate system) expressed in terms of θ, ϕ.

On § 3.2

3.2 Complete the evaluation of scale factors in Example 1.

3.3 Given a two-dimensional space and an orthogonal curvilinear coordinate system (u, v) with $u = x^2 - y^2$, find (i) a simple choice of $v(x, y)$, (ii) the scale factors for these (u, v). Is v unique? Sketch curves of $u =$ const., $v =$ const. in the x–y plane. Express $\hat{\mathbf{u}}$ and $\hat{\mathbf{v}}$ in terms of $\hat{\mathbf{i}}$ and $\hat{\mathbf{j}}$.

3.4 Justify the scale factors in the spherical polar and the cylindrical coordinates geometrically.

3.5 Establish that the following coordinate system is orthogonal. Also find (i) the scale factors (ii) the line elements (iii) the area elements (iv) the volume element.

(i) Parabolic cylinder coordinates:

$$x = \frac{1}{2}(u^2 - v^2), \quad y = uv, \quad z = w.$$

(ii) Conical coordinates:

$$x = \frac{u}{\alpha}\left[(\alpha^2 - v^2)(\alpha^2 + w^2)\right]^{1/2},$$

$$y = \frac{u}{\beta}\left[(\beta^2 + v^2)(\beta^2 - w^2)\right]^{1/2},$$

$$z = \frac{uvw}{\alpha\beta} \qquad \alpha^2 + \beta^2 = 1.$$

3.6 Given $u = \sqrt{x^2 + y^2}$ and $w = z$, obtain a $v(x, y, z)$ by requiring that (u, v, w) form an orthogonal set of coordinates. How is this coordinate system related to the cylindrical coordinates?

3.7 Show that $u = \sqrt{x^2 + y^2} = \rho$, $w = z$ and $v = \varphi \sin \varphi/4$ forms an orthogonal coordinate system.

3.8 From $\hat{\mathbf{u}}_i \cdot \hat{\mathbf{u}}_j = \delta_{ij}$ deduce the following relation:

$$\frac{\partial \hat{\mathbf{u}}_i}{\partial u_k} \cdot \hat{\mathbf{u}}_j + \hat{\mathbf{u}}_i . \frac{\partial \hat{\mathbf{u}}_j}{\partial u_k} = 0 \qquad i \neq j \qquad\qquad k = 1, 2, 3$$

$$\hat{\mathbf{u}}_i . \frac{\partial \hat{\mathbf{u}}_i}{\partial u_k} = 0 \qquad i = 1, 2, 3 \quad k = 1, 2, 3$$

Hence show that if $\frac{\partial \hat{\mathbf{u}}_i}{\partial u_k} \equiv \sum_m a_{ikm} \hat{\mathbf{u}}_m$, then

$$a_{ikm} + a_{mki} = 0$$

$$a_{iki} \equiv 0 \qquad (i \text{ not summed}) \quad i = 1, 2, 3.$$

Hence show that there are in all 9 independent a's.

3.9 Prove that

$$\nabla \times \hat{\mathbf{e}}_1 = \frac{1}{h_1}\left[\frac{\hat{\mathbf{e}}_2}{h_3}\frac{\partial h_1}{\partial u_3} - \frac{\hat{\mathbf{e}}_3}{h_2}\frac{\partial h_1}{\partial u_2}\right]$$

3.10 (a) Show that the two definitions of $\hat{\mathbf{u}}_i$, viz.,

$$\text{(i)}\ \hat{\mathbf{u}}_i = \frac{1}{h_i}\frac{\partial \mathbf{r}}{\partial u_i}; \qquad \text{(ii)}\ \hat{\mathbf{u}}_i = \frac{\nabla u_i}{|\nabla u_i|} = h_i \nabla u_i$$

agree with other.

(b) Using $\dfrac{\partial^2 \mathbf{r}}{\partial u_i \partial u_j} = \dfrac{\partial^2 \mathbf{r}}{\partial u_j \partial u_i}$ show that

$$\frac{\partial \hat{\mathbf{u}}_i}{\partial u_j} = \frac{\hat{\mathbf{u}}_j}{h_i}\frac{\partial h_j}{\partial u_i} \qquad i \neq j.$$

(c) Show further that

$$\frac{\partial \hat{\mathbf{u}}_i}{\partial u_i} = -\sum_{j \neq i} \frac{\hat{\mathbf{u}}_j}{h_j}\frac{\partial h_i}{\partial u_j}.$$

3.11 Show that the two expressions for h_i (Eq. (3.2) and (3.3) of Sec. 3.2) are equivalent.

3.12 Show that if all the three scale factors are unity everywhere, then the coordinate system is Cartesian.

3.13 Show by explicit construction that the Cartesian coordinate system is not the only orthogonal coordinate system in which the triad of unit vectors has a fixed direction at each point. Show however that by performing a scale transformation along each axis, independent of other coordinates, the coordinate system can be converted to the Cartesian coordinates.

3.14 Suppose two curvilinear orthogonal coordinate systems have the same triad of unit vectors at a given point each throughout the space. Show that the two coordinate systems are related by coordinate transformations of the form

$$\tilde{u} = g(u), \qquad \tilde{v} = f(v), \qquad \tilde{w} = g(w).$$

On § 3.3

3.15 Calculate the differential operators ∇, div, curl and ∇^2 for the parabolic coordinate system of Example 2.

3.16 Prove (i) $\nabla \cdot \mathbf{r} = 3$; (ii) $\nabla \times \mathbf{r} = 0$ using spherical polar coordinates.

3.17 Prove $\nabla^2 \frac{1}{r} = 0 \quad (r \neq 0)$, using a suitable coordinate system.

3.18 Prove $\nabla^2 \ln \rho = 0 \quad (\rho \neq 0)$; where ρ stands for the coordinate in the cylindrical coordinate system.

3.19 Show that for a function $\phi(r)$, $\mathbf{r} \times \nabla\phi = 0$.

3.20 For $\psi(r)$, a function of r only, show that

$$\nabla^2\psi = \frac{1}{r^2}\frac{d}{dr}\left[r^2\frac{d\psi}{dr}\right] = \frac{1}{r}\frac{d^2}{dr^2}(r\psi).$$

3.21 Let $\varphi(\mathbf{r}) = \varphi_P$ be the surface of $\varphi =$ constant through a point P. Let $\varphi(\mathbf{r}) = \varphi_P + \Delta\varphi$ be a neighbouring surface of constant φ. ($\Delta\varphi =$ constant). Let Q be any point on this surface. Let $\overrightarrow{PQ} = \hat{\mathbf{t}}\Delta S$. Show that

$$\nabla\varphi\cdot\hat{\mathbf{t}} = \frac{\Delta\varphi}{\Delta S}$$

and is maximum along the normal to the surface of $\varphi = \varphi_P$. Hence conclude that $\nabla\phi$ is perpendicular to this surface and $|\nabla\varphi| = \left|\frac{\partial\varphi}{\partial n}\right|$, the normal derivative of φ.

4 Linear Vector Spaces – I

Linear vector spaces is a topic of primary importance in Physics. The reader is already familiar with the usual three-dimensional (henceforth referred to as 3-D) vector space, say, of position vectors. Its importance in Physics is obvious from the fact that the space in which we live has three dimensions and from the fact that many physical quantities such as force, electric field, velocity, etc. are vectors in the 3-D vector space. Another place where linear vector spaces and linear operators on vector spaces occur in a prominent way is Quantum Mechanics. There the space of wavefunctions (representing states of a physical system) is a linear vector space and the physical observables are represented by linear (hermitian) operators operating on it. In fact, Quantum Mechanics is best formulated in the language of abstract state vectors; from this abstract formulation then, its various concrete formulations/representations can be derived (e.g. wave mechanics, matrix mechanics, Schrodinger, Heisenberg and interaction pictures, etc.). This summarizes a few important reasons why linear vector spaces play a major role in Physics.

In this chapter we focus our attention on abstract linear vector spaces and their basic properties. We leave the topic of linear operators on vector spaces to the next chapter.

4.1 Definition and Examples

The linear vector space we shall deal with generally is an abstract one. This means that the elements of it will not necessarily be identified with any of the familiar entities, though such identification can be made in specific contexts. This allows for a very general discussion in a compact way. Physics, of course, eventually deals with concrete entities. These can be dealt with as special cases of the general discussion.

Consider a set of objects S; each of these objects will be denoted by letters in angular brackets such as $|x\rangle$, $|y\rangle$, $|z\rangle$ $\cdots$, etc. They will be called the abstract vectors. On this set of objects, two operations are defined: (i) addition, and (ii) multiplication by an arbitrary complex numbe

(i) Addition of $|y\rangle$ to $|x\rangle$, both in S, yields another abstract object denoted by $[|x\rangle + |y\rangle]$.

(ii) Multiplication of $|x\rangle$ in S by a complex number α yields another abstract object $\alpha|x\rangle$.

These two operations have to satisfy certain properties given below. In the following, $|x\rangle$, $|y\rangle$, $|z\rangle \in S$ and $\alpha, \beta \in \mathbb{C}$ (read as: $|x\rangle$, $|y\rangle$, $|z\rangle$ belong to the set S and α and β belong to the set of all complex numbers denoted by $\mathbb{C}$).

(1) Commutative law for addition: $|x\rangle + |y\rangle = |y\rangle + |x\rangle$
(2) Associative law for addition: $(|x\rangle + |y\rangle) + |z\rangle = |x\rangle + (|y\rangle + |z\rangle)$
(3) $1|x\rangle = |x\rangle$
(4) Associative law for multiplication $\alpha \cdot (\beta|x\rangle) = (\alpha\beta) \cdot |x\rangle$
(5) Distributive law with respect to addition of complex number

$$(\alpha + \beta)|x\rangle = \alpha|x\rangle + \beta|x\rangle$$

(6) Distributive law with respect to addition of objects in S:

$$\alpha(|x\rangle + |y\rangle) = \alpha|x\rangle + \alpha|y\rangle$$

Note: In the above, for example in (2), the property of associativity of addition is defined only if $(|x\rangle + |y\rangle)$ and $(|y\rangle + |z\rangle)$ are objects in S (a similar thing is also true for (4), (5) and (6)). But a linear vector space is, as we see below, such a set that these assumptions are true by its very definition.

Now, we define the abstract linear vector space (LVS).

Definition: A set S of abstract objects $|x\rangle, |y\rangle, \cdots$ with addition and multiplication by a scalar defined on it is a linear vector space if:

(a) For every $|x\rangle, |y\rangle \in S$; $|x\rangle + |y\rangle \in S$ also.
(b) For every $|x\rangle \in S$ and $\alpha \in \mathbb{C}$, $\alpha|x\rangle \in S$ also.
(c) $\exists$ a null element $|0\rangle \in S$ such that for every $|x\rangle \in S$, $|x\rangle + |0\rangle = |x\rangle$.
(d) For every $|x\rangle \in S$ there exists a "reversed vector" $|x'\rangle \in S$ such that $|x\rangle + |x'\rangle = |0\rangle$.

We shall illustrate these defining properties with the help of concrete examples.

Example 1 Let S be the set of all complex numbers. This set S is a concrete realization of an abstract LVS where every complex number z is regarded as representing an abstract vector $|z\rangle$. We denote this by $|z\rangle \longleftrightarrow z$. We now give the definition of two operation on S. The operation of addition of two vectors here reduces simply to the usual addition of complex numbers which yields another complex number, i.e., another vector in S. The null element is $|0\rangle \longleftrightarrow 0$ and for every $|z\rangle \in S$, $|z'\rangle = |-z\rangle$ is the "reversed vector". (Note in passing that both the null vector of S and the reversed vector for a given $|z\rangle$ are seen to be unique here. As shown later, this is also generally true.) The operation of multiplication of a vector $|z\rangle \longleftrightarrow z$ by a complex number α yields $\alpha|z\rangle \longleftrightarrow |\alpha z\rangle \in S$. All the properties of addition and multiplication by complex number postulated earlier and the defining properties of a LVS are easily verified. (*Note:* In this example, complex numbers happen to play a dual role of both vectors and complex scalars.)

Example 2 Let S = set of all complex n-column vectors. Here we identify

$$|x\rangle \longleftrightarrow X \equiv \begin{bmatrix} x^1 \\ x^2 \\ \vdots \\ x^n \end{bmatrix}$$

where $x^1, x^2, \cdots x^n$ are arbitrary complex numbers. Addition on the set S is the usual matrix addition. Multiplication by complex number takes $|x\rangle$ into $\alpha|x\rangle \longleftrightarrow \alpha X$. Here the null element is easily seen to be $|0\rangle \longleftrightarrow \begin{bmatrix} 0 \\ 0 \\ \vdots \\ 0 \end{bmatrix}$, and the reversed vector for $|x\rangle$ is $|x'\rangle \longleftrightarrow -X$. (Both are, again, unique.) All the relevant properties are easily verified and S is seen to be a LVS.

4.2 Real Linear Vector Space

While giving the definition of the linear vector space, one of the operations introduced on the vectors of the set S was the multiplication of a vector by an arbitrary complex number. We may, however, instead restrict this operation to multiplication of a vector by *real* numbers only. We can then use everywhere only real numbers in defining the properties of (a) addition, (b) multiplication by a number and (c) in the definition of a linear vector space. The vector space we obtain this way is a *real* linear vector space. The usual 3-D vector space is of course the most familiar example of such a real linear vector space because all vectors therein have only real components. For obvious reasons, real linear vector spaces are by themselves quite important in Physics.

4.3 Uniqueness of Null and Reversed Vectors

In the defining properties (c) and (d) of the vector space, we assumed the existence of a null vector $|0\rangle$ in LVS and a reversed vector $|x'\rangle$ for each $|x\rangle$ in LVS. We did not specify that the null and reversed vectors were unique. The defining properties seem, a priori, to leave these possibilities open. However, as shown below, the defining properties themselves do imply the uniqueness of both the null and the reversed vectors. We shall give the proof of this, not so much because the proofs are important from the point of view of Physics, but to illustrate the abstract procedures used in such proofs. One can indeed, as is often done, assume the existence of unique null and reversed vectors in the definitions themselves; but this is strictly not necessary if one is to give the minimal defining properties for LVS.

Theorem 4.1 For every $|x\rangle \in V$, there is a corresponding unique "reversed vector" $|x'\rangle$.

Proof: Let, if possible, there be two distinct reversed vectors $|x'\rangle$ and $|x''\rangle$ for a given $|x\rangle \in V$. Then

$$|x\rangle + |x'\rangle = |0\rangle \qquad \text{a null vector in } V, \tag{4.1}$$

or

$$|x\rangle + |x''\rangle = |0'\rangle. \tag{4.2}$$

Here $|0'\rangle$ is possibly another null vector in V. It may (or may not) differ from $|0\rangle$.

(Note that we are yet to prove the uniqueness of null vector in V. So we cannot assume that the right-hand sides of the two equations (4.1) and (4.2) are necessarily equal.)

Add $|x'\rangle$ to Eq. (4.2).

$$|x'\rangle + (|x\rangle + |x''\rangle) = |x'\rangle + |0'\rangle \tag{4.3}$$

Now, by using associativity of addition on the left-hand side and the definition of null vector on the right-hand side, we have

$$(|x'\rangle + |x\rangle) + |x''\rangle = |x'\rangle. \tag{4.4}$$

Now, by using commutativity of addition and Eq. (4.1)

$$|x'\rangle + |x\rangle = |x\rangle + |x'\rangle = |0\rangle. \tag{4.5}$$

We substitute (4.5) in (4.4) and use $|0\rangle + |x''\rangle = |x''\rangle$ which follows from definition of a null vector. Then we have

$$|x''\rangle = |x'\rangle$$

proving the uniqueness of the reversed vector for any given $|x\rangle \in S$.

Theorem 4.2 The null vector of a vector space S is unique.

Proof: Let, if possible, there be two distinct null vectors $|0\rangle$ and $|0'\rangle$ in S.

Consider some $|x\rangle \in S$. Then there exists a reversed vector $|x'\rangle \in S$ such that $|x\rangle + |x'\rangle$ is a null vector, say $|0\rangle$. (We could proceed equally well assuming it is $|0'\rangle$).

Then,

$$|x\rangle + |x'\rangle = |0\rangle. \tag{4.6}$$

Now, by definition of a null vector $|0'\rangle$,

$$|x\rangle + |0'\rangle = |x\rangle. \tag{4.7}$$

Add $|x'\rangle$ to both sides of (4.7) to get

$$(|x\rangle + |0'\rangle) + |x'\rangle = |x\rangle + |x'\rangle.$$

Then, we use associativity of addition and (4.6)

$$(|x\rangle + |x'\rangle) + |0'\rangle = |0\rangle.$$

That is, $|0\rangle + |0'\rangle = |0\rangle$.

But $|0\rangle$ is a null vector and for every $|x\rangle \in S$, $|0\rangle + |x\rangle = |x\rangle$. Hence the left-hand side is $|0'\rangle$. Hence,

$$|0'\rangle = |0\rangle$$

thus proving the uniqueness of the null vector in S.

Note: The null vector $|0\rangle$ (which is now known to be unique) has the similar familiar algebraic properties of number zero in that (i) $\alpha|0\rangle = |0\rangle$ for any finite α, (ii) $|a\rangle + |0\rangle = |0\rangle = |0\rangle + |a\rangle$. In view of this we shall replace $|0\rangle$ by zero for brevity and as a matter of convention. Here an equation such as $|a\rangle + |b\rangle = |0\rangle$ will be replaced by $|a\rangle + |b\rangle = 0$.

Theorem 4.3 For every $|x\rangle \in V, 0|x\rangle = |0\rangle$.

Proof: From the properties of multiplication by scalar

$$\begin{aligned} |x\rangle &= 1 \cdot |x\rangle \\ &= (1 + 0)|x\rangle \\ &= 1 \cdot |x\rangle + 0|x\rangle \\ &= |x\rangle + 0|x\rangle \end{aligned}$$

where distributivity has been used. Add the reversed vector $|x'\rangle$ to both sides to obtain (now omitting familiar explanations)

$$|0\rangle = (|x'\rangle + |x\rangle) + 0|x\rangle = |0\rangle + 0|x\rangle = 0|x\rangle.$$

Solved example 4.1 Show that the set of continuous complex valued functions $f(x)$ defined for $a < x < b$ forms a LVS. What is the null vector? Is it unique? Is the reversed vector unique?

Solution: First of all, on the set of functions S under consideration, the operation of addition of two functions $f_1(x)$ and $f_2(x)$ to yield a new function $f_1 + f_2$ is defined; so is the multiplication of a function $f(x)$ by a complex number α to yield $\alpha f(x)$. These operations so defined satisfy the six properties listed in Section. 4.

Moreover,

(i) Given two functions $f_1(x)$ and $f_2(x)$ in S, $f_1(x) + f_2(x)$ is a continuous function defined on $a < x < b$ and hence in S.

(ii) Given $f(x) \in S$ and $\alpha \in \mathbb{C}$, $\alpha f(x)$ is a continuous function defined on $a < x < b$ and hence in S.
(iii) The function $n(x) \equiv 0$; $a < x < b$ is an element satisfying the property $f(x)+n(x) = f(x)$ for every $f(x) \in S$.
(iv) Given $f(x) \in S$, the function $\tilde{f}(x) = -f(x)$; $a < x < b$ is a function $\in S$ such that

$$f(x) + \tilde{f}(x) = n(x) = 0 \qquad a < x < b.$$

Hence S is a vector space.

Further, null vector is unique since it must satisfy $f(x)+n(x) = f(x)$, necessarily leading to a unique null vector $n(x) = 0$; $a < x < b$.

The reversed vector for any $f(x) \in S$ is defined by $f(x) + \tilde{f}(x) = n(x) = 0$. This uniquely gives $\tilde{f}(x) = -f(x)$; $a < x < b$. Hence the reversed vector of a given $f(x) \in S$ is unique.

4.4 Scalar Product of Vectors

In this chapter we are dealing with abstract vector spaces. The purpose is to have a single formalism that includes not only the usual 3-D vector space as a special case but also one that includes a wide variety of other concrete realizations as well. If it is to be useful, it must incorporate some of the useful notions used in the context of 3-D vector space in some generalized form. Two of these useful notions are (i) scalar product of two vectors, (ii) metric (measure of distance). In this subsection, we generalize the notion of the scalar product.

Now, in the case of the usual 3-D vector space, scalar product of two vector **A** and **B** is defined by giving directly a way to evaluate it. The vector $|x\rangle$, $|y\rangle$ in an abstract vector space are not known "concrete entities" unlike **A** and **B** in the 3-D vector space. Their scalar product, a complex number denoted by $\langle y|x\rangle$, must be defined in an abstract manner by ascribing (imposing) certain defining properties to it.

So, what are the properties of the scalar product $\mathbf{A} \cdot \mathbf{B}$ in the 3-D vector space that are desirable, so that they may be preserved in its generalization? Some of these are:

(i) Scalar product of **A** and **B** taken in different orders, viz. $\mathbf{A} \cdot \mathbf{B}$ and $\mathbf{B} \cdot \mathbf{A}$ are not independent entities.
(ii) $\mathbf{A} \cdot \mathbf{A} > 0$ unless $\mathbf{A} = \mathbf{0}$, the null vector.
(iii) $\mathbf{A} \cdot \mathbf{0} = 0$.
(iv) The distributive laws: $\mathbf{A} \cdot (\alpha\mathbf{B} + \beta\mathbf{C}) = \alpha\mathbf{A} \cdot \mathbf{B} + \beta\mathbf{A} \cdot \mathbf{C}$.

(Of course, regarding property (i) we have a stronger constraint viz. $\mathbf{A} \cdot \mathbf{B} = \mathbf{B} \cdot \mathbf{A}$; but as can be shown (see Problem. 4.14), it cannot be maintained in the complex generalization.)

Using (i)–(iv) as guidelines, we go on to the general definition of the scalar product.

4.4.1 Definition of scalar product of two vectors

Given any ordered pair $(|x\rangle, |y\rangle)$ of vectors in a linear vector space V, we associate a (generally) complex number denoted by $\langle y|x\rangle$ with it called the scalar product of $|x\rangle$ with $|y\rangle$. It is defined by the following properties:

(i) $\langle x|y\rangle = \langle y|x\rangle^*$

(ii) If $|z\rangle = \alpha|x\rangle + \beta|y\rangle$, then for any $|w\rangle \in V$,

$$\langle w|z\rangle = \alpha\langle w|x\rangle + \beta\langle w|y\rangle.$$

(iii) $\langle 0|0\rangle = 0$ and

$$\langle x|x\rangle > 0 \text{ unless } |x\rangle = |0\rangle.$$

We make several observations on the above definition:

- Firstly, note that unlike in the usual 3-dimensional vector space (with $\mathbf{x}\cdot\mathbf{y} = \mathbf{y}\cdot\mathbf{x}$), $\langle x|y\rangle$ and $\langle y|x\rangle$ may not be equal in general and $\langle x|y\rangle$ is generally a complex number. In the special case when $\langle x|y\rangle$ is real, it is symmetric in $|x\rangle$ and $|y\rangle$. We indeed need an LVS with a complex scalar product in Quantum Mechanics where the wavefunction $\psi(x)$ is looked upon, in Dirac's formalism, as $\langle x|\psi\rangle$, and this is required to be complex in general.
- Secondly, from (i) $\langle x|x\rangle$ is always real and from (iii) it is positive semidefinite (i.e., ≥ 0). $\sqrt{\langle x|x\rangle}$ is called the length of the vector $|x\rangle$.
- If $|z\rangle = \alpha|x\rangle + \beta|y\rangle$, then

  $$\langle z|w\rangle = \langle w|z\rangle^* = \alpha^*\langle x|w\rangle + \beta^*\langle y|z\rangle.$$

  Note the appearance of complex conjugates in the above expression as opposed to the defining property (ii) above. Thus $\langle y|x\rangle$ is said to be linear in $|x\rangle$ and *antilinear* in $|y\rangle$.
- Note that even though the vector space is abstract so that the abstract vector $|x\rangle \in S$ may have no recognizable concrete identity, the scalar product $\langle y|x\rangle$ is always a good old familiar complex number.

Finally, note that the above defining properties may not in general fix *uniquely* the scalar product on a given vector space (e.g., see Problem 4.6 and the solved example/at the end of this section).

Examples

(i) Let V be the vector space of all complex numbers defined earlier with identification $|z\rangle \longleftrightarrow z$. Then $\langle z_1|z_2\rangle \equiv z_1^* z_2$ defines a scalar product. Properties (i) and (ii) are easily verified. Property (iii) also holds because

$$\langle z|z\rangle = z^*z \geq 0 \text{ and } \langle z|z\rangle = 0 \text{ iff } z = 0, \text{ i.e., } |z\rangle = |0\rangle.$$

(ii) Let V be the vector space of complex n-column vectors with identification

$$|x\rangle \longleftrightarrow X = \begin{bmatrix} x^1 \\ x^2 \\ \cdot \\ \cdot \\ x^n \end{bmatrix}$$

Then $\langle y|x\rangle \equiv Y^\dagger X$

defines a scalar product on V. Again properties (i) and (ii) are easily verified (note $X^\dagger Y = (Y^\dagger X)^*$) and

$$\langle x|x\rangle = X^\dagger X = \sum_{m=1}^{n} |x^m|^2 \geq 0.$$

Further, $\langle x|x\rangle = 0$ iff each $x^n = 0$, i.e., iff $X \equiv \begin{bmatrix} 0 \\ 0 \\ 0 \end{bmatrix}$, or equivalently $|x\rangle = |0\rangle$.

4.4.2 Scalar product for real linear vector spaces

As only real linear combinations of vectors are to be considered here, it is necessary to define a scalar product so that it is real, otherwise complex linear combinations of vectors would naturally appear.

Then the defining property (i) is modified to read

$\langle x|y\rangle = \langle y|x\rangle^* = \langle y|x\rangle$, a real number.

Thus $\langle x|y\rangle$ is always symmetric in the two vectors for real vector spaces. A familiar example is that of a usual 3-D vector space with $\langle x|y\rangle = \mathbf{x} \cdot \mathbf{y} = \mathbf{y} \cdot \mathbf{x} = \langle y|x\rangle$.

4.4.3 Cauchy–Schwartz inequality

A certain inequality follows just from the defining properties of the scalar product. This inequality is related to the familiar triangle inequality for the sum of two 3-D vectors.

This inequality, called the Cauchy–Schwartz inequality, reads:

$$|\langle x|y\rangle|^2 \leq \langle x|x\rangle\langle y|y\rangle \text{ for every } |x\rangle\ |y\rangle \in S.$$

To prove this, we note that this is an inequality and contains two vectors $|x\rangle$ and $|y\rangle$. It may be surmised that it can only arise through the use of the third defining property of the scalar product $\langle z|z\rangle \geq 0$, (an inequality in itself), and that too when applied to the general linear combination of $|x\rangle$ and $|y\rangle$. Therefore, we begin with a vector $|z\rangle$ defined by

$$|z\rangle = |x\rangle - \alpha'|y\rangle,$$

where α' is an arbitrary complex constant. For convenience, we shall re-express $\alpha' \equiv \alpha\langle y|x\rangle$, where α is another arbitrary and, a priori, a complex number. (Here we have assumed that $\langle y|x\rangle \neq 0$. A comment on this, later.) Now the vector $|z\rangle$ viz.,

$$|z\rangle \equiv |x\rangle - \alpha\langle y|x\rangle|y\rangle$$

must have nonnegative norm, i.e., $\langle z|z\rangle \geq 0$ for every value of α. It turns out that for our purpose it will suffice to consider only real α. Then $\langle z|z\rangle \geq 0$ leads to

$$\alpha^2|\langle y|x\rangle|^2\langle y|y\rangle - 2\alpha\langle y|x\rangle\,\langle x|y\rangle + \langle x|x\rangle \geq 0 \tag{4.8}$$

which is valid for every real α. This can be rewritten as

$$\alpha^2 - 2\alpha\frac{1}{\langle y|y\rangle} + \frac{\langle x|x\rangle}{\langle y|y\rangle|\langle y|x\rangle|^2} \geq 0.$$

Completing the square,

$$\left[\alpha - \frac{1}{\langle y|y\rangle}\right]^2 + \left[-\frac{1}{\langle y|y\rangle^2} + \frac{\langle x|x\rangle}{\langle y|y\rangle|\langle y|x\rangle|^2}\right] \geq 0. \tag{4.9}$$

As the first term is always nonnegative, the above is satisfied for all real α only if

$$\frac{\langle x|x\rangle}{\langle y|y\rangle|\langle y|x\rangle|^2} - \frac{1}{\langle y|y\rangle^2} \geq 0.$$

This leads to the desired result:

$$|\langle y|x\rangle|^2 \leq \langle x|x\rangle\,\langle y|y\rangle.$$

We finally note that if $\langle y|x\rangle = 0$ the above proof would not apply, but then the Cauchy–Schwartz inequality is trivially true in such a case. This completes the proof.

For brevity we shall refer to the inequality as the C–S inequality.

Examples

(1) Let V be the vector space of all complex numbers with the scalar product $\langle Z_1|Z_2\rangle \equiv Z_1^*Z_2$ defined earlier. Now $\langle Z_1|Z_1\rangle = |Z_1|^2$, $\langle Z_1|Z_2\rangle = |Z_2|^2$. For C–S inequality to hold we should have

$$|Z_1^*Z_2|^2 \leq |Z_1|^2|Z_2|^2.$$

This is trivially fulfilled (in fact as an equality.)

(2) Let V be the 3-D real vector space with the usual scalar product

$$\langle x|y\rangle = \mathbf{X}\cdot\mathbf{Y} = |\mathbf{x}|\,|\mathbf{y}|\cos\theta$$

C–S inequality then reads:

$$|\mathbf{x}\cdot\mathbf{y}|^2 \le |\mathbf{x}|^2\,|\mathbf{y}|^2$$

which is valid as $\cos^2\theta \le 1$.

(3) Let V be the vector space of all complex n-column vectors with the scalar product defined earlier:

$$\langle x|y\rangle = X^\dagger Y$$

This implies that

$$\langle x|x\rangle = X^\dagger X;\ \langle y|y\rangle = Y^\dagger Y$$

C–S inequality would read

$$|X^\dagger Y|^2 \le (X^\dagger X)\,(Y^\dagger Y)$$

This result is not very familiar generally. To see this in an explicit form, we write

$$X = \begin{bmatrix} x^1 \\ x^2 \\ \cdot \\ \cdot \\ x^n \end{bmatrix}; \qquad Y = \begin{bmatrix} y^1 \\ y^2 \\ \cdot \\ \cdot \\ y^n \end{bmatrix}$$

It reads

$$\left|\sum_i x^i y^i\right|^2 \le \sum_i \left|x^i\right|^2 \sum_j \left|y^j\right|^2.$$

(Exercise: Prove the above result directly.)

Solved example 4.2 Consider the usual real 3-D vector space. Define a scalar product on this space by

$$\langle a|b\rangle = a_1 b_1 + \frac{a_2 b_2}{2} + \frac{a_3 b_3}{3}$$

where $|a\rangle$ and $|b\rangle$ represent two vectors **a** and **b** in the 3-D vector space, and a_i and b_i are their components in some fixed reference frame. Show that this satisfies the three defining properties of a scalar product. Is this definition of the scalar product independent of the choice of reference frame? (Unit vectors are to be taken of length unity according to the *usual* scalar product however.)

Solution: This example may look unusual as one is so accustomed to the usual definition of $\mathbf{a}\cdot\mathbf{b}$, but it illustrates the fact that the defining properties above do *not* uniquely fix the scalar product.

(i) $\langle a|b\rangle = \langle b|a\rangle^*$, a real number.

(ii) Let $|c\rangle = \alpha|a\rangle + \beta|b\rangle$. Then

$$\begin{aligned}\langle d|c\rangle &= d_1(\alpha a_1 + \beta b_1) + \frac{1}{2}d_2(\alpha a_2 + \beta b_2) + \frac{1}{3}d_3(\alpha a_3 + \beta b_3)\\ &= \alpha(d_1 a_1 + \frac{1}{2}d_2 a_2 + \frac{1}{3}d_3 a_3) + \beta(d_1 b_1 + \frac{1}{2}d_2 b_2 + \frac{1}{3}d_3 b_3)\\ &= \alpha\langle d|a\rangle + \beta\langle d|b\rangle\end{aligned}$$

(iii) $\langle 0|0\rangle = 0$ is self evident.

$$\langle a|a\rangle = a_1^2 + \frac{a_2^2}{2} + \frac{a_3^2}{3} = 0$$

$$\text{iff } a_1 = 0 \quad \text{and} \quad a_2 = 0 \quad \text{and} \quad a_3 = 0, \quad \text{i.e., } |a\rangle = |0\rangle.$$

Thus the defining properties of the scalar product are verified.

This definition of scalar product is not independent of the choice of coordinate frame. For example, let $a = (1, 0, 0)$ in one frame. Now rotate the axes so that the y-axis of the new frame coincides with the x-axis of the old frame. Compare $\langle a|a\rangle$ in the two frames.

Solved example 4.3 Consider a real 2-D vector space. Define a scalar product (analogous to Example 4.2) $\langle a|b\rangle = a_1 b_1 + \frac{a_2 b_2}{2}$. Verify C–S inequality for this scalar product.

Solution: We want to verify

$$|\langle a|b\rangle|^2 \le \langle a|a\rangle\,\langle b|b\rangle$$

$$\text{i.e., } \left[a_1 b_1 + \frac{a_2 b_2}{2}\right]^2 \le \left[a_1^2 + \frac{a_2^2}{2}\right]\left[b_1^2 + \frac{b_2^2}{2}\right]$$

$$\text{i.e., } a_1^2 b_1^2 + \frac{a_2^2 b_2^2}{4} + a_1 b_1\, a_2 b_2 \le a_1^2 b_1^2 + \frac{a_2^2 b_2^2}{4} + \frac{a_1^2 b_2^2 + a_2^2 b_1^2}{2}$$

$$\text{i.e., } \frac{1}{2}\left[a_1^2 b_2^2 + a_2^2 b_1^2 - 2a_1 b_1 a_2 b_2\right] \ge 0$$

$$\text{i.e., } \left[a_1 b_2 - a_2 b_1\right]^2 \ge 0,$$

which is self-evident.

4.5 Metric Spaces

The idea of a metric on a given set is a straightforward abstract generalization of the familiar idea of distance as defined in the usual 3-D vector space of position vectors. In this generalization, we follow the same procedure as was followed for the scalar product of two vectors, viz. we shall recount the essential properties of distance in the 3-D vector space and define an abstract notion of metric so as to embody these.

The essential properties of distance between two points A and B in the usual 3-D vector space of position vectors are:

(i) Symmetry of the distance between two points, viz. the distance of A from B is equal to the distance of B from A.
(ii) Positivity of distance between two noncoincident points. (If A and B are distinct, distance between A and B is positive)
(iii) The triangle inequality.

We shall define a metric so as to embody these properties.

While we shall not deal with metric spaces at length in this book, we shall introduce this concept here, its relevance being exhibited because of the result proven below: that every linear vector space with a scalar product is a metric space.

Definition: A set R is called a metric space if for every $a, b \in R$ a real nonnegative number $\rho(a, b)$ can be associated with the following properties:

(i) $\rho(a, b) = \rho(b, a)$ (Symmetry)
(ii) $\rho(a, b) \doteq 0$ iff $a \equiv b$
(iii) $\rho(a, b) + \rho(b, c) \geq \rho(a, c)$ (Triangle inequality)

Note that the metric $\rho(a, b)$ is the measure of a "distance" between two elements of a set R. This set need not be a continuous set nor need the "distance" have any physical meaning analogous to ordinary distance. The above three properties define a metric on R. They, however, do not fix it uniquely. (See, for example, Problems 4.15 for examples of metric on a 2-D vector space and their nonuniqueness for a given LVS, and Problem 4.18.)

Now, we come to the theorem we mentioned at the beginning of this subsection.

Theorem 4.4 A linear vector space V with a scalar product defined on it is a metric space.

Proof: To show this we need to establish the existence of a metric on V. To this end, consider

$$\rho^2(a, b) = \| \, |a\rangle - |b\rangle \|^2 \equiv \{\langle a| - \langle b|\} \{|a\rangle - |b\rangle\}$$
$$= \langle a|a\rangle + \langle b|b\rangle - \langle b|a\rangle - \langle a|b\rangle.$$

We shall show that $\rho(a, b)$ so defined above (the nonnegative square root) satisfies all the defining properties of a metric. Evidently, one can see by inspection that $\rho(a, b) = \rho(b, a)$ is satisfied.

Also, the second property is satisfied because

$$\rho(a,b) = 0 \Rightarrow |a\rangle - |b\rangle = 0 \Rightarrow |a\rangle = |b\rangle.$$

It remains, further, to verify the triangle inequality

$$\rho(a,b) + \rho(b,c) \geq \rho(a,c).$$

We note that the R.H.S $= \rho(a,c) = \|\,|a\rangle - |c\rangle\|$.

Now

$$|a\rangle - |c\rangle = \{|a\rangle - |b\rangle\} + \{|b\rangle - |c\rangle\}.$$

We rewrite this as

$$|\alpha\rangle = |\beta\rangle + |\gamma\rangle$$

in obvious new notations. Now

$$\begin{aligned}
\langle\alpha|\alpha\rangle &= \langle\beta|\beta\rangle + \langle\gamma|\gamma\rangle + \langle\beta|\gamma\rangle + \langle\gamma|\beta\rangle \\
&= \langle\beta|\beta\rangle + \langle\gamma|\gamma\rangle + 2\text{Re}\,\langle\beta|\gamma\rangle \\
&\leq \langle\beta|\beta\rangle + \langle\gamma|\gamma\rangle + 2|\langle\beta|\gamma\rangle| \\
&\leq \langle\beta|\beta\rangle + \langle\gamma|\gamma\rangle + 2\sqrt{\langle\beta|\beta\rangle\,\langle\gamma|\gamma\rangle} \\
&= \left\{\sqrt{\langle\beta|\beta\rangle} + \sqrt{\langle\gamma|\gamma\rangle}\right\}^2
\end{aligned}$$

where we have used $\langle\beta|\gamma\rangle = \langle\gamma|\beta\rangle^*$ in the first step and the C–S inequality in the third. Thus, taking the nonnegative square roots, this now leads us to

$$\sqrt{\langle\alpha|\alpha\rangle} \leq \sqrt{\langle\beta|\beta\rangle} + \sqrt{\langle\gamma|\gamma\rangle}$$

Now we use $\rho(a,c) = \|\,|a\rangle - |c\rangle\| = \|\alpha\| = \sqrt{\langle\alpha|\alpha\rangle}$, etc. Thus, we then have

$$\rho(a,c) \leq \rho(a,b) + \rho(b,c)$$

proving the last property desired of a metric too. Thus, we have proved the existence of a metric on V, showing that it is a metric space.

Solved example 4.4 Consider the vector space of all complex numbers. Define a scalar product on this space by $\langle Z_1|Z_2\rangle = Z_1^* Z_2$. Show explicitly that a metric defined through

$$\rho(Z_1, Z_2) = \|\,|Z_1\rangle - |Z_2\rangle\,\|^2$$

has all the usual properties expected of a metric.

Solution: Here,

$$\begin{aligned}\rho(Z_1, Z_2) &= \{\langle Z_1|Z_1\rangle + \langle Z_2|Z_2\rangle - \langle Z_1|Z_2\rangle - \langle Z_2|Z_1\rangle\}^{1/2} \\ &= \{|Z_1|^2 + |Z_2|^2 - Z_1^* Z_2 - Z_2^* Z_1\}^{1/2} \\ &= |Z_1 - Z_2|.\end{aligned}$$

Evidently,

(i) $\rho(Z_1, Z_2) = \rho(Z_2, Z_1)$ as $|Z_1 - Z_2| = |Z_2 - Z_1|$

(ii) $\rho(Z_1, Z_2) = 0 \Leftrightarrow |Z_1 - Z_2| = 0 \Leftrightarrow Z_1 = Z_2$

$$\Leftrightarrow |Z_1\rangle = |Z_2\rangle.$$

We further want to show

(iii) $\rho(Z_1, Z_2) + \rho(Z_2, Z_3) \geq \rho(Z_1, Z_3)$

i.e., $|Z_1 - Z_2| + |Z_2 - Z_3| \geq |Z_1 - Z_3|$.

This is just the actual triangle inequality on the argand diagram where the triangle has as corners the three points Z_1, Z_2 and Z_3 on the Argand plot.

To see this directly, however, we write the square of R.H.S as

$$\begin{aligned}|Z_1 - Z_3|^2 &= |(Z_1 - Z_2) + (Z_2 - Z_3)|^2 \\ &= |Z_1 - Z_2|^2 + |Z_2 - Z_3|^2 + 2\mathrm{Re}\,[(Z_1 - Z_2)(Z_2 - Z_3)^*].\end{aligned}$$

We write the square of L.H.S. as

$$|Z_1 - Z_2|^2 + |Z_2 - Z_3|^2 + 2\mathrm{Re}|Z_1 - Z_2|\,|Z_2 - Z_3|.$$

Proving the triangle inequality amounts to proving

$$\mathrm{Re}\,(Z_1 - Z_2)(Z_2 - Z_3)^* \leq |Z_1 - Z_2|\,|Z_2 - Z_3|$$

But for any complex number ω, $\mathrm{Re}\,\omega \leq |\omega|$.
Taking $\omega = (Z_1 - Z_2)(Z_2 - Z_3)^*$

$$\begin{aligned}\mathrm{Re}\,(Z_1 - Z_2)(Z_2 - Z_3)^* &\leq |(Z_1 - Z_2)(Z_2 - Z_3)^*| \\ &= |Z_1 - Z_2|\,|Z_2 - Z_3|.\end{aligned}$$

Hence etc.

4.6 Linear Independence of Vectors and Basis for a Vector Space

A linear vector space V (barring a trivial example) contains an infinite number of vectors. If it was necessary to specify all of them to characterize the vector space, it would be a forbidding task. Hence, one seeks a certain minimum number of them which when specified allows one to construct the entire vector space because every vector can then be specified in terms of these. Such a set, to be precisely defined later, is referred to as a basis for V. In constructing such a minimal set the idea of linear independence is needed. For example, such a set must not contain a vector which can be expressed as a linear combination of other vectors in this set—for them it would not be minimal. We express this by saying that the set must consist of "linearly independent" vectors only. The basic idea is that a vector $|x\rangle \in V$ is linearly independent of certain other vectors $|y_1\rangle, |y_2\rangle, \ldots$ etc. if $|x\rangle$ cannot be expressed as a linear combination of others. We now proceed to formulate definition of linear independence in a somewhat different manner which, however is more suitable for formal mathematical purposes.

Definition: Consider N vector $|1\rangle, |2\rangle, \ldots |N\rangle$. They are linearly independent if the relation

$$\sum_{i=1}^{N} a_i |i\rangle = 0$$

necessarily implies $a_i = 0$ for $i = 1, 2, \cdots, N$.

Definition: A set of vector which are not linearly independent is said to be linearly dependent.

While both of these definitions are somewhat indirect, they will be shown to be equivalent to the intutive idea presented before.

Example

(1) In the usual 3-D vector space, $\hat{\mathbf{i}}, \hat{\mathbf{j}}$ and $\hat{\mathbf{k}}$ are linearly independent since

$$a_1\hat{\mathbf{i}} + a_2\hat{\mathbf{j}} + a_3\hat{\mathbf{k}} = 0 \Rightarrow a_1 = 0, a_2 = 0, a_3 = 0.$$

(2) Coplanar vectors are linearly dependent. For example $(2\hat{\mathbf{i}}+3\hat{\mathbf{j}})$, $(4\hat{\mathbf{i}}+5\hat{\mathbf{j}})$ and $(+3\hat{\mathbf{i}}+4\hat{\mathbf{j}})$ are linearly dependent since

$$a_1(2\hat{\mathbf{i}} + 3\hat{\mathbf{j}}) + a_2(4\hat{\mathbf{i}} + 5\hat{\mathbf{j}}) + a_3(+3\hat{\mathbf{i}} + 4\hat{\mathbf{j}}) = 0$$

can be satisfied with say $a_1 = 1 = a_2, a_3 = -2$ (among many other solutions).

Theorem 4.5 If N non-null vectors $|1\rangle, |2\rangle, \cdots, |N\rangle$ are linearly dependent then at least one of them can be expressed as a linear combination of others.

Proof: Consider the relation

$$\sum_{i=1}^{N} a_i |i\rangle = 0$$

As the N vectors are linearly dependent, the above equation has a solution with at least one a_i nonzero. Without loss of generality one may assume that $a_1 \neq 0$. We can then write

$$|1\rangle = \sum_{i=2}^{N} \frac{a_i}{a_1} |i\rangle$$

(As $|1\rangle$ is a non-null vector, at least one coefficient $a_i\,(i = 2, \cdots, N)$ is also nonzero.) Hence we have expressed one vector of this set as a linear combination of others.

Finally, note that the null-vector $|0\rangle$ can be trivially expressed as a linear combination of a set of any M vector as

$$|0\rangle = \sum_{i=1}^{M} c_i |i\rangle \text{ with } c_i = 0.$$

For future use, we introduce the following abbreviations:

LC = Linear combination
LI = Linearly independent
LD = Linearly dependent

4.7 Dimension of a Vector Space

Given a vector space V, one can pick up any non-null vector $|x_1\rangle \in V$. The set $\{|x_1\rangle\}$ is obviously a LI set (prove it from the definition). Take any other non-null vector $|x_2\rangle \in V$ and not proportional to $|x_1\rangle$ (i.e., $|x_2\rangle \neq \alpha|x_1\rangle$ for any α) Then the set $\{|x_1\rangle, |x_2\rangle\}$ is also LI Seek to see if there belongs in V a non-null vector $|x_3\rangle$ which is not a linear combination of $|x_1\rangle$ and $|x_2\rangle$, i.e. $|x_3\rangle \neq \alpha|x_1\rangle + \beta|x_2\rangle$ for any $\alpha, \beta \in \mathbb{C}$. Then the set $\{|x_1\rangle, |x_2\rangle, |x_3\rangle\}$ is LI set. Continue this process. For a given vector space V this process may terminate after n steps (i.e. finite number of steps.) This means two things: (i) One has found a set $B = \{|x_1\rangle, \cdots, |x_n\rangle\}$ of LI vectors and (ii) every vector in V is a linear combination of vectors in B. In the case of an another vector space V', this process may never end, i.e., for V' the total number of this set of LI vectors in V' is not finite. In the first case V is said to be a finite-dimensional vector space, while in the second case V' is said to be infinite-dimensional.

Now consider the finite-dimensional vector space V. Could it happen that the number n of elements of B depends on the manner in which the set is chosen? (Note: We have made specific choices of $|x_1\rangle, |x_2\rangle, \cdots$, hence B does depend on this specific process). Suppose by different routes we choose B_1, B_2 and $B_3, \cdots$ having n_1, n_2 and n_3 elements.

We now consider the maximum value of n over all possible combinations of B. (As will be shown later, all constructions of B lead to the same n, but we do not assume it at this stage). This maximum value, n, will be called the dimension of V.

Theorem 4.6 Decomposition of a vector in terms of a given basis is unique.

Proof: We shall assume the vector space to be finite-dimensional.

Let $\{|e_i\rangle; i = 1, \cdots, n\}$ be a basis for an n-dimensional vector space V. Consider any vector $|x\rangle \in V$.

Let, if possible, there be two distinct expansions of $|x\rangle$ in terms of this basis, viz.

$$|x\rangle = \sum_{i=1}^{n} x_i |e_i\rangle \tag{4.10}$$

and

$$|x\rangle = \sum_{i=1}^{n} x_i' |e_i\rangle. \tag{4.11}$$

Substituting Eq. (4.11) in Eq. (4.10), one obtains

$$0 = \sum_{i=1} (x_i - x_i')|e_i\rangle.$$

But, as $\{|e_i\rangle\}$ constitute a basis, they are linearly independent. Hence from the definition of linear independence,

$$x_i - x_i' = 0 \quad \text{or} \quad x_i = x_i',$$

proving that the two expansions are identical.

Note however that the basis for a vector space V itself need not be unique For example, for the usual 3-D vector space, a basis consisting of any three non-coplanar vectors such as $\{\hat{\mathbf{i}}+\hat{\mathbf{j}}, \hat{\mathbf{i}}-2\hat{\mathbf{j}}, \hat{\mathbf{i}}+\hat{\mathbf{k}}\}$ could serve as good a basis as $\{\hat{\mathbf{i}}, \hat{\mathbf{j}}, \hat{\mathbf{k}}\}$. The expansion of a given vector in different bases is (generally) distinct. See the result below.

Theorem 4.7 For an n-dimensional vector space V, any set of n LI vectors constitutes a basis.

Proof: Let $\{|e_i\rangle i = 1, 2, \cdots, n\}$ be any set of n LI vectors in V. We want to prove that this can be used as a basis for V. In other words, we want to show that for every $|x\rangle \in V$, $|x\rangle$ can be expanded as a linear combination of $|e_i\rangle; i = 1, \cdots, n$. To show this consider the equation

$$a_o|x\rangle + \sum_{i=1}^{n} a_i |e_i\rangle = 0. \tag{4.12}$$

If the above equation necessarily implied $a_o = 0$ and $a_i = 0, i = 1, \cdots, n$; then this would imply the existence of $(n+1)$ LI vectors $\{|x\rangle, |e_i\rangle, \cdots, |e_n\rangle\}$ in an n-dimensional vector space V, which is impossible. Hence there exists a solution to the above equation for which at least one of $a_o, a_1, \cdots, a_n$ is nonvanishing. Further, a_o cannot be zero. If this were so, we would have

$$\sum_{i=1}^{n} a_i |e_i\rangle = 0 \tag{4.13}$$

with at least one $a_i (i = 1, \cdots, n)$ nonvanishing. But this is impossible as the set $\{|e_i\rangle\}$ is LI. Hence, (as $a_o \neq 0$), dividing out by a_o, one has

$$|x\rangle = -\sum_{i=1}^{n} \frac{a_i}{a_o} |e_i\rangle \tag{4.14}$$

the desired expansion for every $|x\rangle \in V$. Hence the set $\{|e_i\rangle\}$ constitute a basis for V. Further, by Theorem 4.6 the expansion of Eq. (4.14) is unique.

Solved example 4.5 In the real 3-D vector space, verify that the following vectors are LI.

$$A = \hat{\mathbf{i}} + 2\hat{\mathbf{j}} \quad B = -\hat{\mathbf{j}} + \hat{\mathbf{k}} \quad C = \hat{\mathbf{i}} - \hat{\mathbf{j}} + \hat{\mathbf{k}}$$

Show further that the vector $D = 2(\hat{\mathbf{i}} - \hat{\mathbf{k}})$ can be uniquely expanded as a LC of **A**, **B** and **C**.

Solution: To show the LI of **A**, **B**, **C** we consider the equation

$$a_1\mathbf{A} + a_2\mathbf{B} + a_3\mathbf{C} = 0. \tag{4.15}$$

Comparing coefficients of $\hat{\mathbf{i}}$, $\hat{\mathbf{j}}$ and $\hat{\mathbf{k}}$; we obtain

$$a_1 + a_3 = 0; \; 2a_1 - a_2 - a_3 = 0; \; a_2 + a_3 = 0. \tag{4.16}$$

The first and the last equation leads to $a_1 = a_2 = -a_3$. The second equation implies $a_1 = 0$. Thus the only solution to Eq. (4.15) is $a_1 = a_2 = a_3 = 0$. Hence **A**, **B**, **C** are LI.

To expand the given vector **D** as

$$\mathbf{D} = b_1\mathbf{A} + b_2\mathbf{B} + b_3\mathbf{C};$$

we obtain b_1, b_2, b_3 by comparing coefficients of $\hat{\mathbf{i}}, \hat{\mathbf{j}}, \hat{\mathbf{k}}$;

$$\begin{aligned} 2 &= b_1 + b_3, \\ 0 &= 2b_1 - b_2 - b_3, \\ -2 &= b_2 + b_3. \end{aligned}$$

These leads to $b_1 = -1, b_2 = -5, b_3 = 3$, i.e.,

$$\mathbf{D} = -\mathbf{A} - 5\mathbf{B} + 3\mathbf{C}.$$

4.8 Orthonormal Basis

While any set of n LI vectors can be used as a basis, normally the discussion greatly simplifies if the basis chosen is an orthonormal one, i.e., one that satisfies $\langle e_i|e_j\rangle = \delta_{ij}, 1 \leq i \leq j \leq n$ ($n \to \infty$ for an infinite-dimensional vector space). For example, in the expansion for $|a\rangle$ in an orthonormal basis: $|a\rangle = \sum_i a_i|i\rangle$, a_i is simply given by $\langle i|a\rangle$. It is always possible to choose such a basis as seen below.

Below is given a procedure for constructing an orthonormal basis out of any given basis.

Schmidt orthonormalization theorem

Theorem 4.8 Given a set B of m linearly independent vectors $\{|x_1\rangle, |x_2\rangle, \cdots, |x_m\rangle\}$, a set S of mutually orthonormal vectors $\{|e_1\rangle, |e_2\rangle, \cdots, |e_m\rangle\}$ can always be constructed such that each $|e_i\rangle$ is a linear combination of $|x_j\rangle's$.

Proof: The proof proceeds by explicit construction. We let

$$|e_1\rangle = \frac{|x_1\rangle}{\sqrt{\langle x_1|x_1\rangle}}$$

which is clearly normalized. We now seek to construct a vector $|e_2\rangle$ which is a linear combination of $|x_j\rangle$'s and is orthogonal to $|e_1\rangle$. Clearly this can be done in many ways. Now $|e_2\rangle$ must involve at least one new vector from B other than $|x_1\rangle$ or else it could not be orthogonal to $|e_1\rangle$ (i.e. to $|x_1\rangle$). Choose it to be $|x_2\rangle$. We need to construct a vector from $|x_2\rangle$ orthogonal to $|e_1\rangle$. This is done by removing from $|x_2\rangle$ its component "parallel" to $|e_1\rangle$ so that only component "perpendicular" to $|e_1\rangle$ is left. To be precise consider the vector

$$|e_2'\rangle = |x_2\rangle - \langle e_1|x_2\rangle|e_1\rangle$$

If this vector is non-null it can always be normalized, so that

$$|e_2\rangle = \frac{1}{N_2}\{|x_2\rangle - \langle e_1|x_2\rangle|e_1\rangle\}$$

with N_2 so chosen that $\langle e_2|e_2\rangle = 1$.

We then need to show that this vector is non-null and orthogonal to $|e_1\rangle$, which we shall do shortly in a general setting. But before that we shall give a general procedure for construction of $|e_i\rangle$ and then prove by induction that this procedure works.

Suppose one has already constructed a partial orthonormal set $|e_1\rangle, |e_2\rangle, \cdots, |e_i\rangle (1 \leq i < n)$ out of LC of $|x_1\rangle, \cdots, |x_i\rangle$. We then construct $|e_{i+1}\rangle$ by taking $|x_{i+1}\rangle$ and subtracting components of $|x_{i+1}\rangle$ "parallel" to all $|e_1\rangle, |e_2\rangle \cdots, |e_i\rangle$, and finally by normalizing this vector which can be done provided it is non-null. That is, we let

$$|e_{i+1}\rangle = \frac{1}{N_{i+1}}\left\{|x_{i+1}\rangle - \sum_{j=1}^{i}\langle e_j|x_{i+1}\rangle|e_j\rangle\right\}.$$

Firstly, we note that $|e_{i+1}\rangle \neq 0$; for if it were zero the curly bracket which is an LC of $|x_{i+1}\rangle$ and $|e_1\rangle \cdots |e_i\rangle$ would vanish. But the latter are LC of $|x_1\rangle \cdots |x_i\rangle$. Hence the curly bracket is an LC of LI vectors $|x_1\rangle, |x_2\rangle \cdots, |x_{i+1}\rangle$ and therefore cannot vanish. Hence $|e_{i+1}\rangle \neq 0$. Hence N_{i+1} can be chosen so that $\langle e_{i+1}|e_{i+1}\rangle = 1$.

Secondly, $|e_{i+1}\rangle$ so constructed is orthogonal to $|e_1\rangle \cdots |e_i\rangle$. For,

$$\langle e_j|e_{i+1}\rangle = \frac{1}{N_{i+1}}\{\langle e_j|x_{i+1}\rangle - \langle e_j|x_{i+1}\rangle\} = 0 \qquad 1 \leq j \leq i$$

where use has been made of orthonormality of $|e_1\rangle \cdots |e_i\rangle$. Now, for $i = 1$, $|e_1\rangle$ has already been constructed. Hence proof becomes complete by induction.

Solved example 4.6 Given the LI vectors of Example 4.4 ($\mathbf{A} = \hat{\mathbf{i}} + 2\hat{\mathbf{j}}$, $\mathbf{B} = \hat{\mathbf{k}} - \hat{\mathbf{j}}$, $\mathbf{C} = \hat{\mathbf{i}} - \hat{\mathbf{j}} + \hat{\mathbf{k}}$), construct an orthonormal basis by the Schmidt orthogonalization process.

Solution:

$$|e_1\rangle = \frac{\mathbf{A}}{\sqrt{\mathbf{A}\cdot\mathbf{A}}} = \frac{\mathbf{A}}{\sqrt{5}} = \frac{1}{\sqrt{5}}\hat{\mathbf{i}} + \frac{2}{\sqrt{5}}\hat{\mathbf{j}} = \hat{\mathbf{e}}_1.$$

$$|e_2\rangle = \frac{1}{N_2}\{\mathbf{B} - \mathbf{B}\cdot\hat{\mathbf{e}}_1\hat{\mathbf{e}}_1\} = \frac{1}{N_2}\left\{\mathbf{B} - \frac{\mathbf{B}\cdot\mathbf{A}}{\mathbf{A}\cdot\mathbf{A}}\mathbf{A}\right\}$$

$$= \frac{1}{N_2}\left\{\hat{\mathbf{k}} - \hat{\mathbf{j}} - \frac{(-2)}{5}(\hat{\mathbf{i}} + 2\hat{\mathbf{j}})\right\}$$

$$= \frac{1}{N_2}\left\{\frac{2}{5}\hat{\mathbf{i}} - \frac{1}{5}\hat{\mathbf{j}} + \hat{\mathbf{k}}\right\} \equiv \hat{\mathbf{e}}_2.$$

N_2 is fixed from the normalization condition:

$$\langle e_2|e_2\rangle = 1,$$

$$\text{i.e.,}\quad N_2 = \sqrt{\left(\frac{2}{5}\right)^2 + \left(\frac{1}{5}\right)^2 + 1} = \sqrt{\frac{6}{5}},$$

$$|e_3\rangle = \frac{1}{N_3} = \{\mathbf{C} - \mathbf{C}\cdot\hat{\mathbf{e}}_1\hat{\mathbf{e}}_1 - \mathbf{C}\cdot\hat{\mathbf{e}}_2\hat{\mathbf{e}}_2\}$$

$$= \frac{1}{N_3}\left\{\hat{\mathbf{i}} - \hat{\mathbf{j}} + \hat{\mathbf{k}} - \frac{1}{5}(-1)(\hat{\mathbf{i}} + 2\hat{\mathbf{j}}) - \frac{5}{6}\cdot\frac{8}{5}\left[\frac{2}{5}\hat{\mathbf{i}} - \frac{1}{5}\hat{\mathbf{j}} + \hat{\mathbf{k}}\right]\right\}$$

$$= \frac{1}{N_3}\left\{+\frac{2}{3}\hat{\mathbf{i}} - \frac{1}{3}\hat{\mathbf{j}} - \frac{1}{3}\hat{\mathbf{k}}\right\} = \frac{2\hat{\mathbf{i}} - \hat{\mathbf{j}} - \hat{\mathbf{k}}}{3N_3}.$$

N_3 is easily found to be $\sqrt{\frac{2}{3}}$.

(The reader is encouraged to check directly the orthogonality of $|e_i\rangle$'s.)

4.9 Vector Subspaces

We shall now introduce the idea of a subspace of a vector space. The idea is a familiar one with reference to the usual 3-D vector space characterized by the cartesian unit vectors $\hat{\mathbf{i}}, \hat{\mathbf{j}}$ and $\hat{\mathbf{k}}$. Consider now the set of all vectors lying in (or parallel to) x–y plane. This set of vectors, as can be easily seen, is also a vector space and is a two-dimensional one. We thus have in this example a vector space within a vector space. This set of vectors lying in the x–y plane is called a vector subspace of the 3-D vector space of ordinary vectors. Further, one can easily see that any 3-D vector can be decomposed uniquely into a sum of a component in this subspace (parallel to x–y plane) and a component orthogonal to this subspace (here perpendicular to x–y plane).

Other simple examples of vector subspaces of 3-D vector space are (i) the set of all vectors along x-axis, (ii) the set of all vectors parallel to $a\hat{\mathbf{i}} + b\hat{\mathbf{j}}$ ($a, b =$ const.), (iii) set $\{|0\rangle\}$. These are subspaces with dimension 1, 1 and 0 respectively. The dimension of a vector subspace is defined along the same lines as that of a vector space because the former is a vector space. It is the maximum number of LI vectors in this subspace. Evidently, for a finite-dimensional LVS, the dimension of a subspace $\leq$ dimension of the LVS.

We now proceed to give the rigorous definition of the vector subspace.

Subspaces of linear vector space

Let V be a LVS. Let $S \subset V$ be a set of vectors in V satisfying all the requirements of a vector space by itself. Then S is a subspace of V. In other words, the following must be satisfied.

(i) For every $|x\rangle, |y\rangle \in S$, $|x\rangle + |y\rangle \in S$
(ii) For every $|x\rangle \in S$ and any $\alpha \in \mathbb{C}$, $\alpha|x\rangle \in S$

Here, we note that the requirement that the null vector and the reversed vector (whose existence is guaranteed as V is a vector space) should belong to S is no longer an *independent* requirement; but follows from (ii). For, from (ii), setting $\alpha = 0$, if $|x\rangle \in S$, $0|x\rangle = |0\rangle \in S$ also. Further setting $\alpha = -1$, $(-1)|x\rangle$ which is the reversed vector of $|x\rangle \in S$.

Examples

(i) For the 3-D real vector space:

   (a) All vectors lying in x–y plane constitute a subspace. It is a 2-dimensional subspace.

(b) All vectors lying in any given plane constitute a 2-dimensional vector subspace.

(c) All vectors parallel a given straight line (say x-axis) constitute a 1-dimensional subspace.

(ii) Vector space of complex n-column vectors
Set of all vectors of the form

$$\left\{ \begin{bmatrix} \alpha_1 \\ \alpha_2 \\ \vdots \\ \alpha_m \\ 0 \\ 0 \\ 0 \end{bmatrix} \right\} \} n - m \text{ zeros}$$

where $\alpha_1, \alpha_2, \cdots, \alpha_m$ are arbitrary complex numbers, is an m-dimensional vector subspace of the original n-dimensional vector space.

Next suppose one were to consider the question of how to construct all subspaces of a given vector space V. To see how one can do this, consider any set S of vectors belonging to V. Consider all possible complex linear combinations of vectors in S. Let this set be V_1. Clearly $V_1 \subset V$. Further, as is easy to show, V_1 satisfies the requirement for it to be a vector subspace (Prove!). Hence we have obtained a subspace of V. This is formulated as the result:

Theorem 4.9 Set of all linear combinations of any subset of vectors in a given vector space V is a subspace of V.

Dimensionality of a subspace

Let V_1 be a subspace of V. Then, as $V_1 \subset V$, the number of linearly independent vectors in $V_1 \leq$ no. of LI vectors in V. Hence

$$\dim V_1 \leq \dim V$$

In the special case of $\dim V_1 = \dim V$, V_1 is identical with V (prove it: see Problem 4.28). Technically, the definition of a subspace allows for the possibility of having V as a subspace of itself. In this case V is an "improper subspace" of V. In the case

$$\dim V_1 < \dim V,$$

V_1 is said to be a "proper" subspace of V. In the trivial case $V_1 \equiv \{|0\rangle\}$ which is a subspace of V of dimension zero (prove!); V_1 is a trivial subspace of V.

4.10 Direct Sum Decomposition

For dim $V \geq 2$, there are infinitely many subspaces of V. Suppose one wants to describe V in terms of its subspaces. Is there a (or many) minimal set of subspaces $V_1, V_2, \cdots, V_n$ of V so that if information is provided about these, this gives information about entire V? What would be the characteristic of such a set? Clearly it is redundant to have an overlap between $V_1, V_2, \cdots, V_n$; except the unavoidable overlap of $\{|0\rangle\}$ which must be in each subspace. Further $V_1, V_2, \cdots, V_n$ taken together must be able to "cover" the entire vector space (meaning to be made precise.) That is what is done in a direct sum decomposition of a vector space V. To understand the above idea we can give two examples.

Consider the effect of a rotation R on an arbitrary vector

$$V = a_1\hat{\mathbf{e}}_1 + a_2\hat{\mathbf{e}}_2 + a_3\hat{\mathbf{e}}_3$$

in the real 3-D vector space. We assume for the present that the effect of $\mathbf{V} \rightarrow R\mathbf{V}$ is linear. Then $R\mathbf{V}$ for any $\mathbf{V}$ is known if $R\hat{\mathbf{e}}_1, R\hat{\mathbf{e}}_2, R\hat{\mathbf{e}}_3$ only are given. Hence knowing the effect of the rotation R on vectors of the form $\{\alpha_1\hat{\mathbf{e}}_1\}, \{\alpha_2\hat{\mathbf{e}}_2\}$ and $\{\alpha_3\hat{\mathbf{e}}_3\}$ *only*, i.e. on the three one-dimensional subspaces only, one knows the effect on any $\mathbf{V}$. Similarly for the spin-1/2 system, expectation value of spin $\langle\psi|\mathbf{S}|\psi\rangle$ for any $|\psi\rangle$ is known once the following matrix elements of $\mathbf{S}$: $\langle\uparrow|\mathbf{S}|\uparrow\rangle$, $\langle\downarrow|\mathbf{S}|\downarrow\rangle$ and $\langle\uparrow|\mathbf{S}|\downarrow\rangle$ are known.

We now give a definition which will make the above idea precise.

Definition: Let V be a vector space. Consider subspaces of V, viz. $V_1, V_2, \cdots, V_n$ such that

(i) $V_1 \cap V_j = \{|0\rangle\}$

(ii) Every vector in V can be uniquely decomposed as a sum over vectors belonging to $V_1, V_2, \cdots, V_n$ each, i.e., for every $|x\rangle \in V$, there exist $|x_i\rangle \in V_i$ $i = 1, 2, \cdots, n$ such that

$$|x\rangle = |x_1\rangle + |x_2\rangle + \cdots + |x_n\rangle$$

and this is a unique decomposition of $|x\rangle$.

Then we say that V is a direct sum of $V_1, \cdots, V_n$ and express this as

$$V = V_1 \oplus V_2 \oplus \cdots \oplus V_n.$$

Examples

(i) Let V be the real 3-D vector space.

Let V_1 = set of all vectors along the x-axis.

V_2' = set of all vectors parallel to the $y-z$ plane.

Now, $V_1 \cap V_2' = \{|0\rangle\}$ as $\mathbf{0}$ is the only vector common to V_1 and V_2'. Further it is easily seen that every vector in V can be uniquely decomposed as a vector along x-axis and a vector parallel to $y-z$ plane. Hence

$$V = V_1 \oplus V_2'.$$

(ii) V and V_1 are the same as in Example 4.16
Let V_2 = set of all vectors along y-axis
V_3 = set of all vectors along z-axis.
As is easily seen,

$$V_2 = V_1 \oplus V_2 \oplus V_3.$$

(Also note: $V_2' = V_2 \oplus V_3$).

Theorem 4.10 Let for a n-dimensional vector space V, $V = V_1 \oplus V_2 \oplus \cdots \oplus V_n$. Then the sum of dimensions of subspaces $V_1, V_2, \cdots, V_n$ is equal to n, i.e.,

$$\sum_{i=1}^{n} \dim V_i = \dim V = n.$$

Proof: Consider the subspaces $V_1, V_2, \cdots, V_n$ with a basis for each. Firstly, one notes that the set B of all the basis vectors of $V_1, V_2, \cdots, V_n$ taken together must be a LI set. If they are not LI, then a relation of the sort

$$|x_1\rangle + |x_2\rangle + \cdots |x_n\rangle = |0\rangle \tag{4.17}$$

would hold where $|x_i\rangle \in V_i$ and at least one of $|x_i\rangle$ is a non-null vector. But we also have the decomposition of $|0\rangle$

$$|0\rangle + |0\rangle + \cdots + |0\rangle = |0\rangle \tag{4.18}$$

where first $|0\rangle \in V_1$, second $|0\rangle \in V_2$, etc., and this decomposition must be unique. Now Eq. (4.17) and (4.18) would give distinct decompositions of $|0\rangle$, a contradiction, as this decomposition must be unique.

Now the total number of vectors in the bases for all of $V_1, \cdots, V_n$ is $\sum_{i=1}^{n} \dim V_i$, and as these are LI vectors in an n-dim vector space, this number must be less than or equal to n:

$$\sum_{i=1}^{n} \dim V_i \leq n. \tag{4.19}$$

On the other hand, every vector can be expanded in terms of its "components" in $V_1, V_2, \cdots, V_n$, and in turn the component in V_i can be expanded in terms of the basis for V_i. Thus every vector in V can be expanded in terms of the set B. Hence B must contain at least n LI vectors.

$$\sum_{i=1}^{n} \dim V_i \geq n. \tag{4.20}$$

Equations (4.19) and (4.20) lead us to the desired result

$$\sum_{i=1}^{n} \dim V_i = n. \tag{v}$$

Theorem 4.11 If a finite-dimensional vector space V is a direct sum of two subspaces V_1 and V_2, i.e., $V = V_1 \oplus V_2$, then the union of any basis B_1 of V_1 with any basis B_2 of V_2 is a basis of V.

Proof: For any $|x\rangle \in V$, we have

$$|x\rangle = |x_1\rangle + |x_2\rangle$$

with $|x_1\rangle \in V_1$ and $|x_2\rangle \in V_2$. Let us choose any basis $B_1 = \{|b_1\rangle, \cdots |b_s\rangle\}$ in V_1. Then as $|x_1\rangle \in V_1$ it can be expressed as

$$|x_1\rangle = \sum_{i=1}^{s} \beta_i |b_i\rangle.$$

Similarly, with a basis $B_2 = \{|c_1\rangle, \cdots |c_q\rangle\}$ for V_2

$$|x_2\rangle = \sum_{j=1}^{q} \gamma_j |c_j\rangle.$$

Thus every $|x\rangle \in V$ can be expressed as

$$|x\rangle = \sum_{i=1}^{s} \beta_i |b_i\rangle + \sum_{i=1}^{q} \gamma_j |c_j\rangle,$$

i.e., in terms of the elements of the set $B_1 \cup B_2$. Further, all vectors in B_1 and B_2 taken together form an LI set. (Proof along the lines of the previous theorem.) Hence $B_1 \cup B_2$ is a basis for V.

Solved example 4.7 Consider a set V of all polynomials in x of degree 9 or less. Show that this is a vector space. Show that set of all even polynomials in x of degree 9 or less, V_1, is a vector subspace of this space. Find dimensions of V_1 and V. Find the orthogonal subspace V_2 and verify the direct sum decomposition $V = V_1 \oplus V_2$ explicitly. Find the dimensionality of V_2. Show that there is an isomorphism between V_1 and V_2.

Solution: The proof that V is a vector space is left to the reader (see Problem 4.1).

Let $f(x) \in V_1$. Then, $f(x)$ is of the form $f(x) = a_0 + a_2x^2 + a_4x^4 + a_6x^6 + a_8x^8$.

i) $f(x) \in V_1, g(x) \in V_1 \Rightarrow f(x) + g(x) \in V_1$ for it is an even-ordered polynomial in x of degree 9 or less.
ii) $f(x) \in V_1 \Rightarrow \alpha f(x) \in V_1$ for any $\alpha \in \mathbb{C}$, as is evident.

Hence (i) and (ii) $\Rightarrow V_1$ is a subspace of V.

Every element of V is of the form $f(x) = \sum_{n=0}^{9} b_n x^n$. Each element can be uniquely written as an LC of 10 elements of V: $1, x, x^2, \cdots, x^9$. These 10 elements are linearly independent for a relation

$$\sum_{r=0}^{9} C_r x^r = 0 \quad -\infty < x < \infty,$$

which necessarily implies $C_r = 0, r = 0, \cdots, 9$. Hence V is a 10-dimensional vector space.

Every element of V_1 is expressible as an LC of a basis consisting of the elements $1, x^2, x^4, x^6, x^8$, (which are already known to be LI). Hence dim $V_1 = 5$.

Clearly the dimension of the orthogonal subspace is

$$\dim V - \dim V_1 = 5.$$

We want to find subspace orthogonal to V_1. Consider any element of V, say $f(x)$. To find the orthogonal subspace, one removes from $f(x)$ the "component" parallel to V_1, i.e., one substracts out completely the even part of $f(x)$. One is then left with the odd polynomial $f_{\text{odd}} = f(x) - f_{\text{even}}$. This set of odd polynomials of degree 9 or less is also (similarly) a vector subspace V_2. Further, every polynomial $f(x)$ can uniquely be decomposed as $f(x) = f_{\text{even}} + f_{\text{odd}}$. Moreover, $V_1 \cap V_2 = \{|0\rangle\}$, i.e. only $n(x) = 0$ is both an even and odd polynomial. Hence

$$V = V_1 \oplus V_2$$

There is an isomorphism between V_1 and V_2. Consider $f(x) \in V_1 : f(x) = a_o + a_2x^2 + a_4x^4 + a_6x^6 + a_8x^8$. Then $xf(x) \in V_2$. Moreover, this correspondence is one-to-one. For if $g(x) \in V_2$, $g(x)$ can be expressed uniquely as $xf(x)$. This correspondence is preserved under vector addition and multiplication by a scalar. Such one-to-one correspondence is called isomorphism.

Problems

On § 4.1 and 4.2

4.1 Show that the set of all polynomials in x of degree $\leq n$ form a vector space. (Assume that the coefficients are arbitrary complex numbers.) What is the null vector? What is the reversed vector?

4.2 Show that there is a one-to-one correspondence between elements of the two vector spaces of (i) that in Problem 4.1 above with $n \to (n-1)$ and (ii) that in the Example 2 of § 4.1, and that this correspondence is preserved under vector space operations. Then they are said to be "isomorphic".

4.3 Show that the set $m \times n$ matrices with complex elements forms a vector space. Find the null and the reversed vectors. Can you make a correspondence between this vector space and a vector space of (mn)-tuples?

4.4 Show that if the coefficients of polynomials in set of Problem 4.1 are restricted to real numbers, this set is a real vector space and that it is isomorphic to vector space of n-dimensional vectors. Prove a similar result for the set of matrices in Problem 4.3 whose elements are real.

On § 4.3 and 4.4

4.5 Show that the null vector is orthogonal to every vector in the vector space V, and moreover, it is the only such vector in V.

4.6 In Example 1 of § 4.1, could one define inner product as Re $(Z_1^* Z_2)$? Could one define the inner product as Re $(Z_1^* Z_2)$ for a real linear vector space whose elements are all complex numbers?

4.7 Given vectors $|a\rangle$ and $|b\rangle$ with $\langle b|a\rangle \neq 0$, show that the vector $|b\rangle$ can be resolved as $|b\rangle = \alpha|a\rangle + |d\rangle$, where $|a\rangle$ and $|d\rangle$ are orthogonal.

4.8 Let V be the set of all continuous complex valued functions of x defined over $0 \leq x \leq 1$. Show that this forms a vector space. Show further that

$$\langle g|f\rangle \equiv \int_0^1 g^*(x) f(x) dx$$

defines a scalar product on V. (Here $|f\rangle \longleftrightarrow f(x)$.) Hence show that $P_1(x) = x$ and $P_3(x) = (5x^3 - 3x)/2$ are orthogonal vectors.

4.9 Let V the vector space V of all 2×2 complex matrices. Show that $\langle g|f\rangle = \frac{1}{2}\mathrm{Tr}\,(G^\dagger F)$ defines a scalar product on V. Here $|f\rangle \longleftrightarrow F$, where F is a complex 2×2 matrix. Show further that vectors represented by matrices

$$|e_1\rangle = \begin{bmatrix} 0 & 1 \\ 1 & 0 \end{bmatrix}, |e_2\rangle = \begin{bmatrix} 0 & -1 \\ 1 & 0 \end{bmatrix},$$

$$|e_3\rangle = \begin{bmatrix} 1 & 0 \\ 0 & -1 \end{bmatrix} \quad \text{and} \quad |e_0\rangle = \begin{bmatrix} 1 & 0 \\ 0 & 1 \end{bmatrix},$$

satisfy the orthonormality condition $\langle e_i|e_j\rangle = \delta_{ij}$.

4.10 In Problem 4.9 show that every element on V can be written as $|a\rangle = \sum_{i=0}^{3} a_i|e_i\rangle$; and further that

$$\langle b|a\rangle = \sum_{i=0}^{3} b_i^* a_i.$$

4.11 Consider the vector space V in Problem 4.8. Show that

$$\left|\int_0^1 g^*(x)\, f(x)dx\right|^2 \leq \int_0^1 |g|^2 dx \int_0^1 |f^2| dx.$$

Verify the above for $g(x) = 2ix + 5,\ f(x) = x^2 - 1$.

4.12 Show that the C–S inequality for two non-null vectors becomes an equality $|\langle b|a\rangle|^2 = \langle a|a\rangle\langle b|b\rangle$ iff $|b\rangle$ and $|a\rangle$ are multiples of each other, i.e., $|b\rangle = \alpha|a\rangle$. Reconcile the result in Example (1) of § 4.4.2 (that for the vector space under consideration there, C–S inequality is always an equality) with this result.

4.13 For three vectors $|\alpha\rangle, |\beta\rangle, |\gamma\rangle$ belonging to V, let $|\gamma\rangle = |\alpha\rangle - |\beta\rangle$. Show then that (the triangle inequality) $||\,|\gamma\rangle|| \geq |\,||\,|\alpha\rangle|| - ||\,|\beta\rangle||\,| = |\sqrt{\langle\alpha|\alpha\rangle} - \sqrt{\langle\beta|\beta\rangle}|$ Show further that the equality is achieved for non-null $|\alpha\rangle$ and $|\beta\rangle$ only when $|\beta\rangle = \xi|\alpha\rangle$ with ξ real and positive.

4.14 In the defining properties of a scalar product $\langle y|x\rangle$, we required $\langle y|x\rangle = \langle x|y\rangle^*$. Justify this by showing that this condition is indeed necessary if every scalar product (norm) of the form $\langle w|w\rangle$ is to be real.
[Hint: Take $|w\rangle = |x\rangle + |y\rangle$].

On § 4.5

4.15 To see that the metric on a vector space is not unique, consider the vector space of complex doublets: $V = \left\{\begin{pmatrix} \alpha_1 \\ \alpha_2 \end{pmatrix}\right\}$. Let

(i) $\rho_1(|\alpha\rangle, |\beta\rangle) = ||\,|\alpha\rangle - |\beta\rangle||$

(ii) $\rho_2(|\alpha\rangle, |\beta\rangle) = |\alpha_1 - \beta_1| + |\alpha_2 - \beta_2|$

(a) Show that ρ_1 is an acceptable definition of a metric (using the theorem in § 4.5).

(b) Show that ρ_2 is an acceptable metric. (You may use: $|z + w| \leq |z| + |w|$).

4.16 Let A and B be two 2×2 complex matrices belonging to the vector space of all 2×2 complex matrices. Using the metric in terms of the scalar product defined in Problem 4.9, show that

$$\rho(A, B) = \sum_j \sum_i (a_{ji} - b_{ji})^* (a_{ji} - b_{ji})$$

From this compute $\rho(\sigma_1, \sigma_2)$ where $\sigma_1 = \begin{bmatrix} 0 & 1 \\ 1 & 0 \end{bmatrix}$ and $\sigma_2 = \begin{bmatrix} 0 & -i \\ i & 0 \end{bmatrix}$ are the first two Pauli matrices.

4.17 Let $|\alpha\rangle$ and $|\beta\rangle$ be two non-null vectors in a real vector space. Find the shortest vector of the form $|\alpha\rangle + x|\beta\rangle$. Is it orthogonal to $|\beta\rangle$? Explain geometrically.

4.18 Consider the 3-D Euclidean space. Let r_{ab} denote the geometrical distance between points a and b. Find out the values of n for which the following definition of ρ is acceptable as a metric:

$$\rho(a, b) = r_{ab}^n$$

On § 4.6 and 4.7

4.19 Show that a set of mutually orthogonal vectors is always linearly independent. Is the converse always true?

4.20 If the vectors $|1\rangle, |2\rangle, \cdots, |N\rangle$ are linearly independent, show that the vector $|N+1\rangle$ is an LC of $|1\rangle, \cdots, |N\rangle$ if the vectors $|1\rangle, |2\rangle, \cdots, |N+1\rangle$ are LD.

4.21 Show that the vector space consisting of the set 2×2 complex matrices (defined earlier) over a complex field (i.e. coefficients in LCs can be complex) is 4-dimensional. Show further that the same set over a real field is a real vector space (i.e., coefficients in LC must be real), but now it is 8-dimensional. In each case, write down an expression for an arbitrary matrix $\begin{bmatrix} Z_1 & Z_2 \\ Z_3 & Z_4 \end{bmatrix}$ in terms of a set of chosen LI vectors. Are these expansions unique?

4.22 Find the dimension of the vector space consisting of $n \times n$ complex traceless diagonal matrices over a complex field.

On § 4.8

4.23 Apply the Schmidt orthogonalization procedure to the following set of vectors to obtain an orthonormal basis:

$$\mathbf{A} = \hat{\mathbf{i}} + \hat{\mathbf{j}};\ \mathbf{B} = 2\hat{\mathbf{i}} + 4\hat{\mathbf{j}};\ \mathbf{C} = \hat{\mathbf{i}} + 2\hat{\mathbf{j}} + 3\hat{\mathbf{k}}$$

4.24 Consider the vector space of real polynomials of x of degree 2 or less defined over the interval $-1 \le x \le 1$. Find the dimension of this space. Construct an orthonormal basis for this space. Is it unique? Can you find a basis in which all the basis elements have distinct degrees? Do this exercise for the following definition of scalar product:

(a) $\langle g|f\rangle \equiv \int_{-1}^{1} g^*(x) f(x) dx$

(b) $\langle g|f\rangle \equiv \int_{-1}^{1} (x^2 - 1) g^*(x) f(x) dx$

4.25 Consider any real LVS V of dimension n. Consider also, $V^{(n)}$ the n-dimensional Euclidean vector space. Show that a one-to-one correspondence between vectors of V and $V^{(n)}$ can be made that remains valid under vector space operations (of addition of vectors and multiplication by a scalar). Is there a unique way to establish the correspondence?

4.26 Let V be an n-dimensional LVS. Let $\{|x_1\rangle, |x_2\rangle \cdots |x_m\rangle\}$ be an orthogonal set. Show that it is possible to choose $|x_{m+1}\rangle, \cdots |x_n\rangle \in V$ such that the set $\{|x_1\rangle, \cdots, |x_n\rangle\}$ is orthonormal and a basis for V.

On § 4.9 and 4.10

4.27 Verify that the set $\{|0\rangle\}$ is a subspace of any vector space V. What is its dimension?

4.28 Consider a subspace V_1 of a finite-dimensional vector space V. Let $\dim V_1 = \dim V$. Show that $V_1 = V$.

4.29 Let V_1 and V_2 be two subspaces of V. Show that $V_1 \cap V_2$ is also a subspace of V. Is $V_1 \cup V_2$ necessarily a subspace and, if not, under what conditions can it be a subspace?

4.30 Let V be a finite-dimensional LVS and V_1 be its subspace. Show that there exists another subspace V_2 such that $V = V_1 \oplus V_2$.

4.31 S and T are subspaces of an LVS V such that $S \cup T$ is also a subspace. Show that either $S \subset T$ or $T \subset S$.

5 Linear Vector Spaces – II: Linear Operators

In Physics we encounter vector spaces mainly of two sorts: (i) the usual 3-D vector space, (ii) the vector space of states of a Quantum Mechanical system. We shall introduce the idea of an operation on vector spaces with the help of both of these.

Consider an arbitrary vector $\mathbf{x}$ drawn as an arrow with its tail coinciding with the origin. Consider a rotation of this arrow about some arbitrary axis passing through the origin. As a result of this rotation, the arrow will now represent another vector $\mathbf{x}'$. This is generally true for any $\mathbf{x}$. Thus the result of a given rotation (called R) on a vector $\mathbf{x}$ is to yield another vector in the same vector space. We may represent this symbolically by $\mathbf{x}' \equiv R\mathbf{x}$. Thus, we say that the rotation R is an operation on the vector space V in that it takes one from one element of V to another (generally distinct) element of V.

A similar example could be given for the vector space V of the state functions of a quantum mechanical system. Consider a quantum mechanical system whose state is represented by $|a(0)\rangle$ at $t = 0$, $|a(0)\rangle \in V$, the vector space of states of this system. Let, as time evolves, this state be represented by $|a(t)\rangle \in V$. Then time evolution has taken $|a(0)\rangle \in V$ to another vector $|a(t)\rangle \in V$, and can be represented as an operator on V. These examples allow us to construct a general definition of a linear operator.

5.1 Definition

Consider a linear vector space V. An operator A on this vector space acts on the vectors belonging to V to yield (possibly) another vector in V and it is completely defined by giving the rule for its action:

$$A|x\rangle = |y\rangle$$

for all $|x\rangle \in S \subset V$. Here $|y\rangle \in V$. The set of vectors S belonging to V on which the operation of A is defined is called the domain of A and the set of vectors $|y\rangle$ so generated from all $|x\rangle \in S$ is called the range of A.

Example Consider the three-dimensional vector space V_3 consisting of all real polynomials of degree two in x. A typical vector is given by $|a\rangle = a_o + a_1 x + a_2 x^2$. Consider the operator $A = \frac{d}{dx}$. Operation of A on every element of V_3 is defined. Further $A(a_o + a_1 x + a_2 x^2) = a_1 + 2a_2 x \in V_3$. The range of A is however not V, but is the two-dimensional subspace V_2 of linear polynomials in x. Thus, here the domain of A is V_3 and the range of A is V_2.

We shall assume henceforth that the domain of A is V and its range belongs to V (unless otherwise specified). In particular, we cannot construct a matrix representation of A on V unless its domain is V itself.

Definition: An operator A is a linear operator if for every $|x\rangle, |y\rangle \in S$ and every $\alpha, \beta \in \mathbb{C}$,

$$A[\alpha|x\rangle + \beta|y\rangle] = \alpha A|x\rangle + \beta A|y\rangle.$$

Example The operator $A = \frac{d}{dx}$ in the above example is linear.

Henceforth, we shall assume that we are dealing with *linear operators* unless otherwise specified. We shall denote it by LO.

5.1.1 Algebra of linear operators

(i) We define two operators A and B to be equal iff their operation on a vector in V gives identical results, and this is true for every vector in V, i.e.,

$$A = B \quad \text{iff} \quad A|x\rangle = B|x\rangle \quad \text{for every} \quad |x\rangle \in V.$$

(ii) Addition of two operators A and B: Given two linear operators A and B, their addition denoted by $A + B$ is defined as that operator whose action on V is defined by

$$(A + B)|x\rangle = A|x\rangle + B|x\rangle \quad \text{for every} \quad |x\rangle \in V$$
$$(= B|x\rangle + A|x\rangle).$$

Note that on account of last equation $A + B = B + A$.

(iii) Multiplication of two operators A and B: Multiplication of B followed by A, denoted by $A \cdot B$, is that operator whose action on V is defined by

$$A \cdot B|x\rangle = A(B|x\rangle).$$

In general, since there is no reason to expect $A(B|x\rangle) = B(A|x\rangle)$, $A \cdot B$ and $B \cdot A$ may be distinct operators, i.e., generally $A \cdot B \neq B \cdot A$. This is expressed by saying that A and B generally may not commute.

(iv) Powers of A: For a positive integer m, the operator A^m for a given linear operator A is defined as

$$A^m|x\rangle = \underbrace{A[A[\ldots A|x\rangle]\ldots]}_{m} \quad \text{for every} \quad |x\rangle \in V.$$

Exercise Show that $A + B$ and AB are linear operators.

Examples

(1) Consider the usual 3-D vector space. Consider an operation R_1 that corresponds to rotation through $\pi/2$ about the z-axis. Its action on unit vectors is

$$R_1\hat{\mathbf{i}} = \hat{\mathbf{j}} \qquad R_1\hat{\mathbf{j}} = -\hat{\mathbf{i}} \qquad R_1(\hat{\mathbf{k}}) = \hat{\mathbf{k}}.$$

In fact its action on any $\mathbf{A} = a_1\hat{\mathbf{i}} + a_2\hat{\mathbf{j}} + a_3\hat{\mathbf{k}}$ is

$$R_1\mathbf{A} = (-a_2\hat{\mathbf{i}} + a_1\hat{\mathbf{j}} + a_3\hat{\mathbf{k}}) \equiv \mathbf{B}.$$

R_1 is a linear operator as is easily seen by visual inspection. Both the range and domain of R_1 are the entire vector space.
We can introduce another rotation operator R_2 corresponding to rotation through $-\pi/2$ about x-axis. Its operation on unit vectors is given by

$$R_2\hat{\mathbf{i}} = \hat{\mathbf{i}} \qquad R_2\hat{\mathbf{j}} = -\hat{\mathbf{k}} \qquad R_2\hat{\mathbf{k}} = \hat{\mathbf{j}}$$

$$R_2\mathbf{A} = a_1\hat{\mathbf{i}} + a_3\hat{\mathbf{j}} - a_2\hat{\mathbf{k}}.$$

The operator $R_1 \cdot R_2$ is defined by computing the action of R_2 followed by that of R_1 on all unit vectors. Thus $R_1 \cdot R_2\hat{\mathbf{i}} = R_1(\hat{\mathbf{i}}) = \hat{\mathbf{j}}$; $R_1 \cdot R_2\hat{\mathbf{j}} = -\hat{\mathbf{k}}$; $R_1 \cdot R_2\hat{\mathbf{k}} = -\hat{\mathbf{i}}$; $R_1 \cdot R_2\mathbf{A} = -a_3\hat{\mathbf{i}} + a_i\hat{\mathbf{j}} - a_2\hat{\mathbf{k}}$ and $R_1 \cdot R_2$ is moreover a linear operator.
(It is not difficult to see that $R_1 \cdot R_2$ corresponds to a rotation about a line parallel to the vector $(\hat{\mathbf{i}} + \hat{\mathbf{j}} - \hat{\mathbf{k}})$. Note that $R_1 \cdot R_2$ leaves the vector $(\hat{\mathbf{i}} + \hat{\mathbf{j}} - \hat{\mathbf{k}}$ unchanged.)
Now we calculate $R_2 \cdot R_1$.

$$R_2 \cdot R_1\hat{\mathbf{i}} = R_2\hat{\mathbf{j}} = -\hat{\mathbf{k}}; \qquad R_2 \cdot R_1\hat{\mathbf{j}} = -\hat{\mathbf{i}}; \qquad R_2 \cdot R_1\hat{\mathbf{k}} = \hat{\mathbf{j}};$$

$$R_2 \cdot R_1\mathbf{A} = -a_2\hat{\mathbf{i}} + a_3\hat{\mathbf{j}} - a_1\hat{\mathbf{k}}.$$

Clearly $R_1 \cdot R_2 \neq R_2 \cdot R_1$.
Further $R_1^2 = R_1 \cdot R_1$ is clearly seen to be

$$R_1^2\hat{\mathbf{i}} = R_1 \cdot (R_1\hat{\mathbf{i}}) = R_1\hat{\mathbf{j}} = -\hat{\mathbf{i}}; \qquad R_1^2\hat{\mathbf{j}} = -\hat{\mathbf{j}}; \qquad R_1^2\hat{\mathbf{k}} = \hat{\mathbf{k}}.$$

Thus, R_1^2 represents rotation about z-axis through π.

(2) Consider the linear vector space V_3 of all quadratic polynomials. Let $A = \frac{d}{dx}$ as before. Also let $B = x\frac{d^2}{dx^2}$. Then,

$$A + B = \frac{d}{dx} + x\frac{d^2}{dx^2},$$

$$AB = \frac{d}{dx}\left[x\frac{d^2}{dx^2}\right] = \frac{d^2}{dx^2} + x\frac{d^3}{dx^3},$$

$$BA = x\frac{d^2}{dx^2}\frac{d}{dx} = x\frac{d^3}{dx^3}.$$

Note that $AB \neq BA$.

Further, $A^3 = \frac{d^3}{dx^3}$ and for every $|a\rangle \in V$, $A^3|a\rangle = 0$. As we shall see, A^3 will coincide with the null operator.

5.1.2 Minimal definition of a linear operator

While operator A is generally defined by giving its action on every vector in its domain, in the case of linear operators on a subspace V_1 of V (V_1 may possibly coincide with V), its action is known completely once its action on each of the *basis vectors* of V_1 is given.

For, let $|e_1\rangle, |e_2\rangle \ldots$ be a basis for V_1. Now consider any arbitrary vector $|x\rangle \in V_1$. It can be expressed in terms of the basis as $|x\rangle = \sum x_i|e_i\rangle$. Then, using the definition of LI vectors,

$$A|x\rangle = A\sum x_i|e_i\rangle = \sum_i x_i A|e_i\rangle,$$

which is known completely if all of $A|e_i\rangle$ are known*. And this is valid for every $|x\rangle \in V_1$. Thus the linear operator A is completely defined if $A|e_i\rangle (i = 1, 2, \ldots)$ are all given, i.e., by knowing the action of A only on all basis vectors of its domain subspace V_1. Hence the definitions in Section 5.1 need refer only to the basis vector of V rather than each vector in V.

5.1.3 Action of an arbitrary LO on the null vector

Let A be any arbitrary linear operator. Consider any vector $|x\rangle \in V$. Consider the equation

$$|0\rangle = 0|x\rangle.$$

Operating on the above equation by A and using linearity we obtain

$$\begin{aligned} A|0\rangle &= A[0|x\rangle], \\ &= 0A|x\rangle, && (A\alpha|x\rangle = \alpha A|x\rangle) \\ &= |0\rangle. && (0|y\rangle = |0\rangle) \end{aligned}$$

Thus any aribtrary linear operator acting on a null vector yields always the null vector.

5.1.4 Special operators

The null operator: An operator O such that $O|x\rangle = |O\rangle$ for every $|x\rangle \in V$ is called the null operator.

The identity operator: The identity operator I is defined by

$$I|x\rangle = |x\rangle \quad \forall |x\rangle \in V.$$

*For an infinite-dimensional LVS V_1, convergence of series such as the one above has been assumed henceforth.

5.2 Inverse of an Operator

Consider any linear operator A. If there exists a linear operator X such that

$$XA = I,$$

X is called a left inverse of A; and we write $X = A_\ell^{-1}$.

Similarly, for a linear operator A, if there exists an LO Y such that $AY \doteq I$, then Y is called a right inverse of A and we write $Y = A_r^{-1}$.

Note: For a given A,

(i) left inverse may not exist;
(ii) if it exists it may not be unique;
(iii) right inverse may not exist;
(iv) right inverse, if it exists, may not be unique; and
(v) left inverse and right inverse could be a priori distinct even if both exist.

This situation, true in general, may be found confusing at first sight because of its unfamiliarity. The student is familiar with inverse in the case of a (finite) matrix where the inverse, if it exists, is unique and further the left and the right inverses are identical. We shall see later why this must be so for (finite-dimensional) matrices and indeed for finite-dimensional vector spaces. But this need hold for an arbitrary (infinite-dimensional) vector space. To see this consider the following example.

Example Consider an infinite-dimensional vector space spanned by an orthonormal basis vectors $|e_1\rangle, |e_2\rangle, \cdots, |e_n\rangle \cdots$. Let an LO A be defined by

$$A|e_n\rangle = |e_{n-1}\rangle, \qquad n = 2, 3, \cdots$$
$$A|e_1\rangle = 0.$$

(a) Does a left inverse exist? Supposing it existed, operate by it on the last equation

$$A_\ell^{-1}\, A|e_1\rangle = A|0\rangle = |0\rangle,$$
$$\text{i.e.,} \quad I|e_1\rangle = |e_1\rangle = |0\rangle.$$

But $|e_1\rangle$ being of unit norm cannot equal the null vector (which has a zero norm). Hence we reach a contradiction. Thus, the left inverse does not exist

(b) Does a right inverse exist? Answer is given by an explicit construction. Define an operator B by

$$B|e_n\rangle = |e_{n+1}\rangle \qquad n = 1, 2, 3, \cdots$$

Operate on the last equation by A (on the left):

$$AB|e_n\rangle = A|e_{n+1}\rangle = |e_n\rangle \quad n = 1, 2, 3, \cdots$$

Hence $AB = I$; or, $B = A_r^{-1}$.
Thus the right inverse does exist.

(c) Is the right inverse unique?
Again we answer by giving an explicit construction, Define C by

$$C|e_n\rangle = |e_{n+1}\rangle + \alpha_n|e_1\rangle \quad (\alpha_n = \text{an arbitrary complex number}).$$

Then,

$$\begin{aligned} AC|e_n\rangle &= A|e_{n+1}\rangle + |0\rangle \\ &= |e_n\rangle \qquad\qquad n = 1, 2, \cdots \end{aligned}$$

Hence $AC = I$, i.e., C is a right inverse. Now for different $\alpha_n's$, we get different operators C and thus a continuous set of infinite number of right inverses is obtained.

The above exemplifies the complex situation that generally holds in (infinite-dimensional) vector spaces. In Physics an operator has a physical meaning, hence the above complexity would seem unphysical. Fortunately, in the situations one commonly encounters, the state of affairs is much simpler. This is made possible because of the three results two of which we prove now.

Theorem 5.1 If for a given A both a left inverse A_ℓ^{-1} and a right inverse A_r^{-1} exist, then they are unique and identical.

Proof: Here we have $A_\ell^{-1}A = I = A\,A_r^{-1}$. Now

$$A_\ell^{-1} = A_\ell^{-1}I = A_\ell^{-1}(A\,A_r^{-1}) = (A_\ell^{-1}A)A_r^{-1} = IA_r^{-1} = A_r^{-1}.$$

Therefore the given A_ℓ^{-1} and the given A_r^{-1} are equal. If there was another left inverse $A_\ell^{-1'}$ then by the same argument as above

$$A_\ell^{-1'} = A_r^{-1}.$$

Thus $A_\ell^{-1'} = A_\ell^{-1}$ proving the uniqueness of the left inverse. Further, if there was another right inverse $A_r^{-1'}$ then by the same argument as above

$$A_\ell^{-1} = A_r^{-1'}.$$

Hence $A_r^{-1} = A_r^{-1'}$, proving the uniqueness of right inverse.

Theorem 5.2 If A_ℓ^{-1} is unique, then it is also the unique right inverse.

Proof: Consider

$$A(A_\ell^{-1}A) = AI = A = IA \tag{5.1}$$

i.e., $$(A\,A_\ell^{-1} - I)A = 0. \tag{5.2}$$

Also $$A_\ell^{-1}A = I. \tag{5.3}$$

Adding (5.2) and (5.3) we obtain

$$(A\,A_\ell^{-1} + A_\ell^{-1} - I)A = I.$$

Thus, a right inverse exists and is given by $A_r^{-1} = AA_\ell^{-1} + A_\ell^{-1} - I$. Thus both A_ℓ^{-1} and A_r^{-1} exist, and by the previous theorem, they are unique and identical.

Comment: On account of this theorem, we may drop, in this case, the subscript and write $A_\ell^{-1} = A_r^{-1} = A^{-1}$, the inverse of A.

5.3 Dual spaces and adjoint operators

5.3.1 Dual vector space

Let V be a LVS. From this LVS we construct another LVS (of the same dimension), V' by the following rule:

(i) For each element $|x\rangle \in V$, we introduce an element *denoted* by $\langle x| \in V'$. (Note that we are using a "different-looking" notation for the vectors in V'.)

(ii) For each element of the form $|z\rangle \equiv [\alpha|x\rangle + \beta|y\rangle] \in V$, we associate it with $\langle z| \equiv [\alpha^*\langle x| + \beta^*\langle y|] \in V'$.

One can go back to the defining properties of LVS in Chapter 4 and verify that the set of vectors $V' \equiv \{\langle a|\}$ so defined, in fact corresponds to an LVS. (See Problem 5.52.)

V' is referred to as the "dual vector-space" of V. We saw that the definition of the scalar product of two vectors $|x\rangle, |z\rangle \in V$, viz. $\langle z|x\rangle$ was *linear* in $|x\rangle$ and *antilinear* in $|z\rangle$. We can alternately look upon the scalar product $\langle z|x\rangle$ as constituted from the vector $\langle z|$ dual to $|z\rangle$ and belonging to V', and the vector $|x\rangle \in V$. Then the defining property (ii) above implies that $\langle z|x\rangle$ is linear in both $\langle z|$ and $|x\rangle$ (see Problem 5.54).

5.3.2 Adjoint of an operator

Let A be an operator. Operator A was defined by giving $A|x\rangle$ for all $|x\rangle \in V$. Equivalently, operator A could also be completely known if $\langle y|A|x\rangle$ is known for all $|x\rangle$ and $|y\rangle \in V$ (see Problem 5.2). $\langle y|A|x\rangle$ are, in Quantum Mechanics, often referred to as the "matrix elements" of A (we shall be precise about it when we introduce matrix representations of A). In terms of these matrix elements, we define a new (related) operator $A^\dagger$, also by giving all its "matrix elements".

$$\langle x|A^\dagger|y\rangle \equiv \langle y|A|x\rangle^* \quad \text{for all} \quad |x\rangle, |y\rangle \in V.$$

The definition is very analogous to the usual definition of adjoint of a finite-dimensional matrix and we will clarify the connection later.

Properties:

(i) $(A^\dagger)^\dagger = A$. Consider

$$\langle y|(A^\dagger)^\dagger|x\rangle = \langle x|A^\dagger|y\rangle^* = [\langle y|A|x\rangle^*]^*$$
$$= \langle y|A|x\rangle \quad \text{for all} \quad |x\rangle, |y\rangle \in V.$$

Hence $(A^\dagger)^\dagger = A$.

(ii) $(A+B)^\dagger = A^\dagger + B^\dagger$ (Prove)

(iii) $(AB)^\dagger = B^\dagger A^\dagger$

5.3.3 Operators expressible as $|x\rangle\langle y|$

In 5.3.1 we introduced the dual of a vector. Dual to $|y\rangle$ is denoted by $\langle y|$. Consider an object denoted by $|x\rangle\langle y|$. We *define* this as an linear operator with the defining property

$$\{|x\rangle\langle y|\}|z\rangle \equiv |x\rangle\langle y|z\rangle = \langle y|z\rangle|x\rangle.$$

Such expressions are often a convenient way of characterizing operators. Later on we shall see that every linear operator can be expressed as a linear combination of such terms.

Next we find the hermitian conjugate of the linear operator $A = |x\rangle\langle y|$. It is defined by the requirement

$$\langle z|A^\dagger|w\rangle = \langle w|A|z\rangle^*.$$

We shall show that this hermitian conjugate is given by $|y\rangle\langle x|$. Call this operator B. We then have to show that for every $|z\rangle$ and $|w\rangle \in V$

$$\langle z|B|w\rangle = \langle z|A^\dagger|w\rangle \equiv \langle w|A|z\rangle^*$$

Now, by definition,

$$A|z\rangle = |x\rangle\ \langle y|z\rangle,$$
$$\langle w|A|z\rangle = \langle w|x\rangle\ \langle y|z\rangle,$$
$$\langle w|A|z\rangle^* = \langle x|w\rangle\ \langle z|y\rangle = \langle z|y\rangle\ \langle x|w\rangle.$$

On the other hand,

$$B|w\rangle = |y\rangle\,\langle x|w\rangle.$$

$$\therefore \quad \langle z|B|w\rangle = \langle z|y\rangle\,\langle x|w\rangle = \langle w|A|z\rangle^* = \langle z|A^\dagger|w\rangle.$$

Hence, etc.

Definition: The equation $\langle x| = \langle y|$ is said to be dual to the equation $|x\rangle = |y\rangle$.

Definition: Following Dirac, we call $|x\rangle$ and $\langle y|$ as the 'ket' and 'bra' vectors so that the quantity $\langle y|x\rangle$ constitutes the bracket.

5.4 Special Linear Operators

Now we shall discuss a few special kinds of linear operators that occur commonly in Physics.

(i) Hermitian operator: An operator H is called hermitian iff $H = H^\dagger$.
Hermitian operators are very important in quantum mechanics as all physical observables are represented by Hermitian operators. We shall study properties of these in more detail later.

(ii) Unitary operator: An operator U is called unitary if a unique U^{-1} exists and is equal to $U^\dagger$, i.e., $UU^\dagger = U^\dagger U = \mathbb{1}$.
An important property of a unitary operator is that it preserves the norm of a vector. By this we mean that $|x\rangle$ and $U|x\rangle \equiv |y\rangle$ have the same norm for any $|x\rangle \in V$. Norm of $|y\rangle$ is $\langle y|y\rangle$ with $\langle y| = \langle x|U^\dagger$. Hence $\langle y|y\rangle = \langle x|U^\dagger U|x\rangle = \langle x|I|x\rangle = \langle x|x\rangle$. This property is very commonly used in Quantum Mechanics, where often it refers to conservation of probability under various physical operations, all of which are represented by unitary (or antinunitary) operators.

(iii) Antihermitian operator: An operator A is called antihermitian iff $A^\dagger = -A$.
Note if H is hermitian then $\pm i\ H$ are antihermitian and if A is antihermitian then $\pm i\ A$ are hermitian. Note that every operator B can be decomposed uniquely in terms of a hermitian and an antihermitian part:

$$B = \frac{B + B^\dagger}{2} + \frac{B - B^\dagger}{2} \equiv H + A.$$

(See Problem 5.56).

(iv) Antilinear operator: Physics generally needs only linear operators but for an important exception. The operator T acting on the quantum mechanical state vectors of a system that represents time reversal operation is an "antilinear" operator. An operator A is antilinear if for every $|x\rangle, |y\rangle \in V$ and for every complex numbers α and β,

$$A[\alpha|x\rangle + \beta|y\rangle] = \alpha^* A|x\rangle + \beta^* A|y\rangle.$$

A is given completely if its action on each of the basis vectors of V is known; provided one uses the above property.

(v) Antiunitary operator: An antilinear operator that is also unitary is called an antiunitary operator. The time reversal operator T mentioned earlier is antiunitary.

5.5 Projection Operators

Until now, we have generally dealt with the entire vector space V. One may, sometimes, be interested only in one of the subspaces, say V'. Now, as we have seen in Section 4.10, every vector in the vector space can be uniquely decomposed into two parts—one belonging to the subspace V' and one orthogonal to this subspace. One may want to project the part of any vector belonging to the subspace V' only. For this, we introduce what is called a *projection operator* P.

For example, consider the usual 3-D vector space V with a typical vector $\mathbf{r} = x\hat{\mathbf{i}} + y\hat{\mathbf{j}} + z\hat{\mathbf{k}}$, and its 2-D subspace V' of vectors parallel to the x–y plane. A projection operator P can project every vector in V into its projection on V', i.e.,

$$P(\hat{\mathbf{r}}) = x\hat{\mathbf{i}} + y\hat{\mathbf{j}} = \mathbf{r}' \in V'.$$

Since a projection operator P projects an arbitrary vector $|x\rangle$ into its projection belonging to the subspace V'; any further operation of P cannot change the vector $P|x\rangle$, for $P|x\rangle$ is already in the subspace V'. Hence $P\,P|x\rangle = P|x\rangle$ for every $|x\rangle \in V$. Hence every projection operator must satisfy $P^2 = P$.

Given below are examples of projection operations.

(1) Consider a vector space spanned by orthonormal vectors $|e_1\rangle, |e_2\rangle, \cdots |e_n\rangle, \cdots$. Consider the operator $P_1 = |e_1\rangle\,\langle e_1|$. P_1 projects any vector $|x\rangle \in V$ into the subspace V_1 spanned by $|e_1\rangle$.
For $P_1|x\rangle = |e_1\rangle\,\langle e_1|x\rangle$, the component of $|x\rangle$ along $|e_1\rangle$ and $P_1|x\rangle \in V_1$, P_1 is a projection operator. For using normalization of $|e_1\rangle$,

$$P_1^2 = |e_1\rangle\langle e_1|e_1\rangle\langle e_1| = |e_1\rangle\langle e_1| = P_1.$$

Also note $P_1^\dagger = |e_1\rangle\langle e_1| = P_1$ so that P_1 is hermitian.

(2) In the example above, consider $P_2 = 1 - P_1$. P_2 projects any arbitrary vector $|x\rangle \in V$ into the orthogonal subspace V_2 spanned by $\{|e_2\rangle, |e_3\rangle, \cdots |e_m\rangle, \cdots\}$, i.e.,

$$P_2|x\rangle = |x\rangle - |e_1\rangle\,\langle e_1|x\rangle = \sum_m |e_m\rangle\,\langle e_m|x\rangle - |e_1\rangle\langle e_1|x\rangle$$

$$= \sum_{m=2} |e_m\rangle\,\langle e_m|x\rangle \in V_2.$$

Note here that V is a direct sum of V_1 and V_2.

$$V = V_1 \oplus V_2,$$
$$\text{and} \quad |x\rangle = P_1|x\rangle + P_2|x\rangle$$

is the unique decomposition of $|x\rangle$ in terms of components $P_1|x\rangle$ and $P_2|x\rangle$ belonging to the subspaces V_1 and V_2. Further, it is easily verified that $P_2^2 = (1 - P_1)^2 = P_2$.

The above discussion of projection operators can be easily generalized to an arbitrary direct sum decomposition of V. For example, consider an n-dimensional LVS V with

$$V = V_1 \oplus V_2 \oplus V_3 \oplus \cdots \oplus V_m.$$

We can define m projection operators $P_1, P_2, \cdots, P_m$ such that $P_i|x\rangle \in V_i$ for every $|x\rangle \in V$. Each operator satisfies $P_i^2 = P_i$.

Further, $P_i P_j$, for $i \neq j$ vanishes. This can be seen easily from the defining property. For any $|x\rangle$, $P_j|x\rangle \in V_j$. P_i acting on any vector projects its component in V_i. $P_i P_j|x\rangle$ must therefore belong both to V_i and V_j; the only such vector is the null vector. Hence the operator $P_i P_j$ acting on any vector $|x\rangle$ yields the null vector and hence is the null operator by definition. Thus,

$$P_i P_j = \delta_{ij} P_i \quad i = 1, 2, \cdots, m.$$

Such a set $\{P_1, \cdots, P_m\}$ is called a set of orthogonal projection operators.

Exercises

(i) Show that a projection operator has no inverse unless $P = I$, the trivial projection operator.

(ii) Show that for an orthogonal set of projection operators $\{P_1, \cdots, P_m\}$, any sum of them (say, $P_1 + P_2 + P_5 + P_7$) is also a projection operator. Identify the vector subspace onto which is projection operator projects.

5.6 Eigenvalues and Eigenvectors

Definition: Consider a linear operator A defined in a linear vector space. Consider the equation

$$A|x\rangle = \lambda|x\rangle.$$

If a non-null vector $|x\rangle \in V$ exists that satisfies the above equation for some (possibly complex) number λ, then $|x\rangle$ is called an eigenvector of A belonging to an eigenvalue λ.

Examples

(1) Let V be the LVS of 2×2 matrices. Let $A = \begin{bmatrix} 0 & 1 \\ 1 & 0 \end{bmatrix}$ be an LO on it. We see easily that

$$A\begin{pmatrix}1\\1\end{pmatrix} = 1\cdot\begin{pmatrix}1\\1\end{pmatrix}$$

$$A\begin{pmatrix}1\\-1\end{pmatrix} = -1\begin{pmatrix}1\\-1\end{pmatrix}$$

Thus $\begin{pmatrix}1\\1\end{pmatrix}$ is an eigenvector of A with eigenvalue 1 and $\begin{pmatrix}1\\-1\end{pmatrix}$ is an eigenvector of A with eigenvalue -1.

Before passing onto the next example, we make some observations with the help of this example.

(i) For $\alpha \neq 0$, $\begin{pmatrix}\alpha\\\alpha\end{pmatrix} = \alpha\begin{pmatrix}1\\1\end{pmatrix}$ is also clearly seen to be eigenvector of A with eigenvalue 1. Thus, the overall (nonzero) normalization of an eigenvector is not fixed by the eigenvalue equation.

(ii) In this example, $AX = \lambda X$ has no non-null solution for X whenever $\lambda \neq 1$ or -1. (Verify). This shows that A has only two eigenvalues. (Later on we shall see that an $n \times n$ matrix has at most n distinct eigenvalues.)

(iii) The eigenvectors $\begin{pmatrix}1\\1\end{pmatrix}$ and $\begin{pmatrix}1\\-1\end{pmatrix}$ corresponding to the distinct eigenvalues are LI.

(2) Let V be the LVS of differentiable complex functions of a single real variable x. Let $A = \frac{d}{dx}$. Consider

$$\frac{d}{dx}e^{ikx} = ik\,e^{ikx}.$$

e^{ikx} is an eigenfunction of A with eigenvalue (ik).

Definition: The set $(\lambda_1, \lambda_2, \cdots, \lambda_n, \cdots)$ of eigenvalues of a linear operator A is called the spectrum of A.

Before proceeding, we shall make a number of simple observations:

(i) Every non-null vector $|x\rangle$ is an eigenvector of the identity operator with eigenvalue 1, since $I|x\rangle = 1 \cdot |x\rangle$.

(ii) The identity operator can have no eigenvalue other than 1.

(iii) If an operator has every (non-null) vector in V as its eigenvector, then it is a multiple of the identity (see Problem 5.60).

(iv) If spectrum of A is $(\lambda_1, \lambda_2, \cdots, \lambda_n, \cdots)$, then the spectrum of A^m is $(\lambda_1^m, \lambda_2^m, \cdots, \lambda_n^m, \cdots)$. For $A|x\rangle = \lambda_1|x\rangle$ implies, by successive operation, $A^m|x\rangle = \lambda_1^m|x\rangle$.

(v) Each eigenvector of A is also eigenvector of A^m. The converse is, however, not true (see Problem 5.25).

5.7 Results on the Eigenvalues and Eigenvectors of Special Operators

5.7.1 Hermitian and unitary operators

In the following we shall prove certain results about special kinds of operators, viz. hermitian operators and unitary operators. Such operators occur naturally in a number of places and are of paramount importance in Quantum Mechanics. In Quantum Mechanics, all observables are represented by hermitian operators. Most transformations in Quantum Mechanics (such as those for space translations, for time translations, for rotations, and one for going from coordinate to momentum basis, etc.) are represented by unitary operators because they preserve the norm of the state vectors (see Problem 5.15). Hermitian and unitary operators occur in other places as well. For example, inertia tensor and polarizability tensor are examples of real hermitian matrices. Transformation matrices between two Cartesian coordinate systems (see for example Eq. (1.11) of Section 1.1.1) are real unitary matrices.

Theorem 5.3

(i) Eigenvalues of a hermitian operator are real and
(ii) eigenvectors of a hermitian operator and corresponding to different eigenvalues are orthogonal.

Proof: Consider the eigenvalue equation:

$$H|x\rangle = \lambda|x\rangle, \tag{5.4}$$

where $|x\rangle$ is an eigenvector corresponding to an eigenvalue λ of H. Taking the dual of the equation (i.e., writing the corresponding vectors in the dual vector space) and using $H^\dagger = H$,

$$\langle x|H = \lambda^*\langle x|. \tag{5.5}$$

Taking scalar product of (5.4) with $\langle x|$ and of (5.5) with $|x\rangle$ and combining,

$$\langle x|H|x\rangle = \lambda\ \langle x|x\rangle = \lambda^*\langle x|x\rangle.$$

As $|x\rangle$ is non-null, the above equation implies $\lambda = \lambda^*$, i.e., λ is real. This proves the first part. To prove the second part, consider another eigenvector $|y\rangle$ with a distinct eigenvalue $\lambda' \neq \lambda$, now known to be real.

$$H|y\rangle = \lambda'|y\rangle. \tag{5.6}$$

Taking dual as before and using $H^\dagger = H$ and $\lambda'^* = \lambda'$,

$$\langle y|H = \lambda'^*\langle y| = \lambda'\langle y|. \tag{5.7}$$

Taking scalar product of (5.4) with $\langle y|$ and (5.7) with $|x\rangle$ and combining, we find

$$\langle y|H|x\rangle = \lambda\langle y|x\rangle = \lambda'\langle y|x\rangle. \tag{5.8}$$

As $\lambda \neq \lambda'$, this leads to $\langle y|x\rangle = 0$, i.e., $|x\rangle$ and $|y\rangle$ are mutually orthogonal.

Comments: The theorem has important implications in Quantum Mechanics. The eigenvalues of hermitian operators representing physical observables are, according to postulates of Quantum Mechanics, the possible results of a measurement of the observable. The above theorem assures that these results are then real, as should the result of any physical measurement be.

Theorem 5.4

(i) Eigenvalues of a unitary operator are unimodular.
(ii) Eigenvectors of a unitary operator corresponding to different eigenvalues are orthogonal.

Proof:

(i) Consider the eigenvalue equation for a unitary operator U:

$$U|x\rangle = \lambda|x\rangle \tag{5.9}$$

and its dual

$$\langle x|U^\dagger = \lambda^*\langle x|. \tag{5.10}$$

Taking scalar product of the bra and ket vectors in (5.9) and (5.10), and using $U^\dagger U = I$,

$$\langle x|U^\dagger U|x\rangle = \langle x|x\rangle = |\lambda|^2\langle x|x\rangle.$$

As $\langle x|x\rangle \neq 0$, this implies $|\lambda|^2 = 1$, i.e. λ is unimodular.

(ii) Now, consider the eigenvalue equation of another distinct eigenvalue λ':

$$U|y\rangle = \lambda'|y\rangle. \tag{5.11}$$

Taking scalar product of (5.10) with (5.11).

$$\langle x|U^\dagger\ U|y\rangle = \langle x|y\rangle = \lambda^*\lambda'\langle x|y\rangle. \tag{5.12}$$

Now using $U^\dagger U = I$ and recalling $\lambda^*\lambda = 1$ we can write (5.12) as

$$\langle x|y\rangle = \lambda^*\lambda\langle x|y\rangle = \lambda^*\lambda'\langle x|y\rangle. \tag{5.13}$$

Hence

$$\lambda^*(\lambda - \lambda')\ \langle x|y\rangle = 0.$$

Now as $\lambda^* \neq 0$ (remember $|\lambda| = 1$) and $\lambda - \lambda' \neq 0$, it follows that

$$\langle x|y\rangle = 0,$$

proving the orthogonality of $|x\rangle$ and $|y\rangle$, viz. the eigenvectors of U with distinct eigenvalues.

5.7.2 Normal operators

We have just seen two instances of operators viz. hermitian and unitary operators which have eigenvectors orthogonal when they refer to distinct eigenvalues. We shall later show that when either kinds of operators have more than one LI eigenvector corresponding to a given eigenvalue, one can choose, using Schmidt orthogonalization process, to make this set of LI eigenvectors mutually orthogonal. It thus becomes possible to construct an orthonormal set of eigenvectors for such operators. In the case of finite-dimensional vector spaces, we shall also show later that this set is complete. These results are of great physical significance. To give examples, consider the inertia tensor $I_{\alpha\beta}$ of a rigid body. This is known to be a real symmetric matrix and is thus a hermitian matrix (for definition of hermitian matrices see Section 5.4). Its eigenvalues are real and called the principle moments. More importantly, its principle axes are those whose directions are given by its eigenvectors, which we know form a mutually orthogonal set (or can be chosen to form so). The same applies to the polarizability tensor of a biaxial crystal. Its principle axes are the principle dielectric axes and these are mutually orthogonal because they lie along the eigenvectors of this tensor.

In Quantum Mechanics, observables are represented by Hermitian operators with a complete set of eigenvectors. This set then can be chosen to be mutually orthogonal. For example, the Hamiltonian (which is the operator corresponding to total energy) is a hermitian operator. In particular its eigenvalues are real, which must be, if they are to correspond to the energies of states. Furthermore, the eigenvectors corresponding to the states of definite energy are all mutually orthogonal.

A mathematical question that naturally arises is what is the most general class of operators that has this property, viz., its eigenvectors are (or can be chosen to be) mutually orthogonal. In this connection we prove a theorem below.

Definition: For two operators A and B defined on V, $[A, B] \equiv AB - BA$ is an operator on V and is called the commutator of A with B.

Definition: An operator is called normal iff $[A, A^\dagger] = 0$.

Theorem 5.5 Eigenvectors of an operator A belonging to distinct eigenvalues are mutually orthogonal if $[A, A^\dagger] = 0$, i.e., A is normal.

Proof: Consider a linear operator A, for which $AA^\dagger = A^\dagger A$. We will first show that for such an LO, every eigenvector of A is also an eigenvector of $A^\dagger$ with eigenvalue complex-conjugated. To see this, consider the eigenvalue equation

$$A|x\rangle = \lambda|x\rangle. \tag{5.14}$$

We wish to show then that, in the present case,

$$A^\dagger|x\rangle = \lambda^*|x\rangle. \tag{5.15}$$

This is equivalent to showing that the vector $A^\dagger|x\rangle - \lambda^*|x\rangle \equiv |y\rangle$ is a null vector. This is done by using the property unique to the null vector $\langle 0|0\rangle = 0$, i.e., a null vector is the only vector with a zero norm. Now,

$$\begin{aligned}\langle y|y\rangle &= \{\langle x|A - \lambda\langle x|\}\,\{A^\dagger|x\rangle - \lambda^*|x\rangle\}\\ &= \langle x|AA^\dagger|x\rangle - \lambda^*\langle x|A|x\rangle - \lambda\langle x|A^\dagger|x\rangle + |\lambda|^2\langle x|x\rangle.\end{aligned}$$

In the first term on the R.H.S. we use $AA^\dagger = A^\dagger A$ and then use

$$\langle x|A^\dagger = \lambda^*\langle x| \tag{5.16}$$

in it and also in the third term [(5.16) is implied by (5.14)]. We use (5.14) in the second term. We then find the four terms mutually cancelling out and thus obtain

$$\langle y|y\rangle = 0.$$

This then implies

$$|y\rangle = 0 = A^\dagger|x\rangle - \lambda^*|x\rangle.$$

Hence, Eq. (5.15) is proven.

Now, consider further a distinct eigenvalue of A, viz. λ', and the corresponding eigenvalue equation:

$$A|x'\rangle = \lambda'|x'\rangle. \tag{5.17}$$

Taking the dual of Eq. (5.15),

$$\langle x|A = \lambda\langle x|. \tag{5.18}$$

Taking scalar product of (5.18) with $|x'\rangle$ and of (5.17) with $\langle x|$, we find

$$\langle x|A|x'\rangle = \lambda\langle x|x'\rangle = \lambda'\langle x|x'\rangle. \tag{5.19}$$

Thus

$$(\lambda - \lambda')\langle x|x'\rangle = 0. \tag{5.20}$$

As $\lambda \neq \lambda'$, $\langle x|x'\rangle = 0$, i.e. the eigenvectors $|x\rangle$ and $|x'\rangle$ of A with distinct eigenvalues λ and λ', are mutually orthogonal. This proves the result.

Comments:

(i) The above theorem states the sufficient condition for orthogonality of eigenvectors corresponding to distinct eigenvalues.

(ii) For a finite-dimensional vector space, the theorem assures that if the eigenvalues are all distinct, there is a complete set of mutually orthogonal eigenvectors.

(iii) In the case of degenerate eigenvalues, the theorem however, does not say that the set of eigenvectors can be chosen to be a complete orthonormal set. However, for finite-dimensional vector space, we shall show later that this is always possible for a normal A. (See comments after Theorem 5.18.)

(iv) In the case of finite-dimensional vector spaces we shall show that A must be normal for this property of orthogonality of eigenvectors to hold. (See Theorem 5.18.)

5.8 Similarity and Unitary Transformations

A given physical system can have a variety of descriptions. To give a simple example, an electric field in space can be described using different choices of coordinate systems, leading to different descriptions. Physical phenomena in relativistic physics can be described by using different inertial frames. In the case of Quantum Mechanical systems, one can describe them using coordinate representation or momentum representation or energy representation. They can alternately be described in the Schrodinger picture or the Heisenberg picture. Going from one physical description to another involves a transformation of the quantities describing the physical state. Such transformations are similar to the similarity transformations defined below. Yet these transformations must be such that they preserve some essential physically observable characteristics. In this section we discuss these transformations, and a special case of these—the unitary transformations.

Definition: Let A and S be linear operators on a linear vector space 'V'. Then the transformation $A \to A' = \mathrm{SAS}^{-1}$ is called a similarity transformation.

Also, $A = S^{-1}A'S = S^{-1}A'(S^{-1})^{-1}$. Thus A is related to A' by a similarity transformation. Thus if A' is similar to A then A is similar to A'.

Theorem 5.6 Under similarity transformation $A \to \mathrm{SAS}^{-1}$, the set of eigenvalues of A remains ulatered. For every eigenvector $|x\rangle$ of A there is an eigenvector $S|x\rangle$ of SAS^{-1} with the same eigenvalue.

Proof:

$$A|x\rangle = \lambda|x\rangle.$$

Now, consider

$$[SAS^{-1}]S|x\rangle = SAI|x\rangle = SA|x\rangle = \lambda S|x\rangle.$$

Thus $S|x\rangle$ is an eigenvector of SAS^{-1} with the same eigenvalue λ. Thus the set of eigenvalues of SAS^{-1} contain the set of eigenvalues of A. But the transformation

$$SAS^{-1} \to A = S^{-1}[SAS^{-1}](S^{-1})^{-1}$$

is also a similarity transformation. Hence the set of eigenvalues of A contain the eigenvalues of SAS^{-1}. Hence, the two operators have an identical set of eigenvalues.

Example Consider the Pauli spin matrices

$$\sigma_x = \begin{bmatrix} 0 & 1 \\ 1 & 0 \end{bmatrix} \qquad \sigma_y = \begin{bmatrix} 0 & -1 \\ i & 0 \end{bmatrix} \qquad \sigma_z = \begin{bmatrix} 1 & 0 \\ 0 & -1 \end{bmatrix}.$$

These constitute operators on V_2, the LVS of 2×1 complex column vectors. They are all related to each other by a similarity transformation.

$$\sigma_x = S_{(1)}\,\sigma_z S_{(1)}^{-1}$$
$$\sigma_y = S_{(2)}\,\sigma_z S_{(2)}^{-1}$$
$$\sigma_x = S_{(3)}\,\sigma_y S_{(3)}^{-1}$$

with

$$S_{(1)} = \frac{1}{\sqrt{2}}\begin{bmatrix} 1 & 1 \\ 1 & -1 \end{bmatrix} = S_{(1)}^{-1},$$
$$S_{(2)} = \frac{1}{\sqrt{2}}\begin{bmatrix} 1 & i \\ i & 1 \end{bmatrix},$$
$$S_{(3)} = S_{(1)}S_{(2)}^{-1},$$

as is easily verified.

The eigenvalues of all the three are identical, viz. 1 and -1. The eigenvectors of σ_z are $\begin{bmatrix} 1 \\ 0 \end{bmatrix}$ and $\begin{bmatrix} 0 \\ 1 \end{bmatrix}$ and hence the eigenvectors of σ_x are $S_{(1)}\begin{bmatrix} 1 \\ 0 \end{bmatrix}$ and $S_{(1)}\begin{bmatrix} 0 \\ 1 \end{bmatrix}$ and similarly for σ_y.

5.9 Matrix Representations

We, now turn to an important topic—both in itself and also from the point of view of Physics—that of matrix representations of vectors in an LVS and of linear operators on a vector space. The LV spaces we have been dealing with have generally been abstract in nature, we did not commit ourselves to any concrete form (except in examples). Now we shall show that an n-dimensional vector space V with a LO defined on it can be "represented" in a concrete form by matrices of appropriate type, to be introduced below. The word representation is used in a detailed mathematical sense: it implies that there will be a one-to-one correspondence between vectors of V and n-column vectors and between linear operators and $n \times n$ matrices, and that this correspondence will be preserved under all the operations of the vector space and of linear operators. Matrix representations give us a convenient way of bringing the abstract concepts of vectors and linear operators closer into a familiar domain of complex matrices. They are very commonly used in Physics. Quantum Mechanics uses them frequently as the physically observable quantities are neither the abstract operators (representing observables) nor the abstract state-vectors, but rather the matrix elements of operators between state vectors.

5.9.1 Construction of a representation

Consider an n-dimensional LVS 'V'. Consider a linear operator A on V. To construct the matrix representation (henceforth MR for short) for the vectors belonging to V and of a linear operator A on V, we choose a basis in V, say $B \equiv \{|e_1\rangle, |e_2\rangle, \cdots, |e_n\rangle\}$. (The basis need not be orthonormal. Also note that, as elaborated later, whatever follows now does depend on a specific choice for the basis.) We, then, expand any vector $|x\rangle \in V$ as

$$|x\rangle = \sum_{i=1}^{n} x^i |e_i\rangle. \tag{5.21}$$

Then we associate with $|x\rangle$ the column vector X given below:

$$|x\rangle \longleftrightarrow X = \begin{pmatrix} x^1 \\ x^2 \\ \vdots \\ x^n \end{pmatrix}$$

X constitutes a matrix representation of $|x\rangle \in V$ (in terms of the chosen basis B).

Now, to construct the MR for the linear operator A we evaluate $A|e_i\rangle$ for $i = 1, 2, \cdots, n$ and write

$$A|e_i\rangle = \sum_{i=1}^{n} a_{ji} |e_j\rangle. \tag{5.22}$$

These define n^2 coefficients $(a_{ji, i, j=1,2,\cdots,n})$. Then we construct a matrix

$$\bar{A} = \begin{bmatrix} a_{11} & a_{12} & a_{1n} \\ \vdots & \vdots & \vdots \\ a_{n1} & a_{n2} & a_{nn} \end{bmatrix}.$$

The $n \times n$ matrix $\bar{A}$ is then the MR of the linear operator A in this basis.

Now, we shall proceed to obtain the MR of a product of two operators say A and B. In doing so, the usual rule for matrix multiplication (normally introduced in an ad-hoc fashion and hence somewhat of a mystery) will be "derived" naturally.

Consider

$$AB|e_i\rangle = A\left\{\sum_j b_{ij}|e_i\rangle\right\} = \sum_j b_{ji}A|e_i\rangle$$

$$= \sum_i b_{ji}\sum_k a_{kj}|e_k\rangle = \sum_{k=1}^{n}\left[\sum_{j=1}^{n} a_{kj}\, b_{ji}\right]|e_k\rangle. \tag{5.23}$$

The quantity in square brackets is normally defined as the $(k, i)^{\text{th}}$ element of the matrix product $\bar{A}\bar{B}$. We, thus, have

$$AB|e_i\rangle = \sum_k (\bar{A}\ \bar{B})_{ki}|e_k\rangle.$$

Comparing this equation with Eq. (5.2) above, we find that the definition of matrix product is chosen in such a manner that the operator AB has an MR $\bar{A}\bar{B}$, i.e., equal to the matrix product of the MR's of A and B.

We also find that the effect of operation of a linear operator A on any vector $|x\rangle \in V$ can be expressed in terms of $\bar{A}$ and X. To see this consider

$$A|x\rangle = A\left\{\sum_i x^i|e_i\rangle\right\} = \sum_i x^i A|e_i\rangle$$

$$= \sum_i x^i \sum_j a_{ji}|e_j\rangle = \sum_j\left[\sum_i a_{ji}x^i\right]|e_j\rangle$$

$$= \sum_j (\bar{A}\ X)_i|e_i\rangle,$$

where $\bar{A}\ X$ is the usual matrix product of a square matrix $\bar{A}$ and a n-column vector X. Thus the vector $A|x\rangle$ has as its MR the column vector $\bar{A}\ X$.

For an n-dimensional abstract vector space, the entire calculations done in it (in terms of LO's and vectors) can be done equivalently in terms of the more familiar matrices. Thus the set of complex n-column vectors is an LVS (which is a representation of V) and the set of $n \times n$ matrices (which is a representation of LO's on V) can be used alternately and equivalently.

Example Consider an LVS 'V' of quadratic real polynomials in x. $f(x) \in V$ implies $f(x) = b_1 + b_2x + b_3x^2$. V is 3-dimensional and there is one–one correspondence between $f(x)$ and the real column vectors of V_3:

$$f(x) \longleftrightarrow \begin{pmatrix} b_1 \\ b_2 \\ b_3 \end{pmatrix}.$$

Here we have chosen $\{1, x, x^2\}$ as the basis for V.

Consider further a linear operator $A = \frac{d}{dx}$. Clearly,

$$A|e_1\rangle = \frac{d}{dx}1 = 0 = |0\rangle, \qquad \langle e_i|A|e_1\rangle = 0$$

$$A|e_2\rangle = \frac{d}{dx}x = 1 = 1|e_1\rangle, \qquad \langle e_i|A|e_2\rangle = \delta_{i1}$$

$$A|e_3\rangle = \frac{d}{dx}x^2 = 2x = 2|e_2\rangle. \qquad \langle e_i|A|e_3\rangle = 2\delta_{i2}$$

Thus MR of A is easily seen to be

$$\bar{A} = \begin{bmatrix} 0 & 1 & 0 \\ 0 & 0 & 2 \\ 0 & 0 & 0 \end{bmatrix}.$$

Thus the equation $\frac{d}{dx}(1 + 5x + 3x^2) = 5 + 6x$ is represented by the matrix equation

$$\begin{bmatrix} 0 & 1 & 0 \\ 0 & 0 & 2 \\ 0 & 0 & 0 \end{bmatrix} \begin{bmatrix} 1 \\ 5 \\ 3 \end{bmatrix} = \begin{bmatrix} 5 \\ 6 \\ 0 \end{bmatrix}.$$

Consider further the operator $B = x\frac{d}{dx}$. It is easily seen that B has matrix representation

$$\bar{B} = \begin{bmatrix} 0 & 0 & 0 \\ 0 & 1 & 0 \\ 0 & 0 & 2 \end{bmatrix}.$$

The equation $AB = \frac{d}{dx}\left(x\frac{d}{dx}\right) = \frac{d}{dx} + x\frac{d^2}{dx^2}$ is represented by the matrix equation

$$\bar{A}\bar{B} = \begin{bmatrix} 0 & 1 & 0 \\ 0 & 0 & 2 \\ 0 & 0 & 0 \end{bmatrix} \begin{bmatrix} 0 & 0 & 0 \\ 0 & 1 & 0 \\ 0 & 0 & 2 \end{bmatrix} = \begin{bmatrix} 0 & 1 & 0 \\ 0 & 0 & 4 \\ 0 & 0 & 0 \end{bmatrix} = \bar{C},$$

$\bar{C}$ being the MR of the operator $\frac{d}{dx} + x\frac{d^2}{dx^2}$ (as is easily verified). (Of course, it may be remarked that this example is only of pedagogical interest as there is no great mathematical advantage in going to an MR here.)

In passing we remark that for an operator A

$$\langle e_j|A^\dagger|e_i\rangle = \langle e_i|A|e_j\rangle^* = a_{ij}^*.$$

Thus the adjoint of A is represented by a matrix $\underline{A}^\dagger$ with the matrix elements

$$(\underline{A}^\dagger)_{ji} = a_{ij}^*.$$

Further, since AB has the MR $\bar{A}\bar{B}$, a unitary operator (which satisfies $U^\dagger U = UU^\dagger = I$) will have an MR $\underline{U}$ satisfying $\underline{U}^\dagger \underline{U} = \underline{U}\,\underline{U}^\dagger = I$. Such a matrix $\underline{U}$ is called a unitary matrix. Similarly a hermitian operator H (satisfying $H = H^\dagger$) will have an MR $\underline{H}$ satisfying $\underline{H} = \underline{H}^\dagger$. Such a matrix $\underline{H}$ is called a hermitian matrix.

A real symmetric matrix X is a special case of a Hermitian matrix since $S = S^T = S^\dagger$ for it. A real orthogonal matrix is a special case of a unitary matrix since $O^{-1} = O^T = O^\dagger$.

5.10 Results on the Eigenvalues and Eigenvectors of Matrices

In this section, we shall prove a number of fundamental results on the eigenvalues and the eigenvectors. Further we shall establish a number of important results regarding certain special types of matrices which will be of great value in Physics, expecially in Quantum Mechanics.

Theorem 5.7 Every $n \times n$ matrix has at least one eigenvector.

Proof: Consider the eigenvalue equation

$$AX = \lambda X. \tag{5.24}$$

We want to show (5.24) will have at least one solution for

$$X \neq \begin{pmatrix} 0 \\ 0 \\ \vdots \\ 0 \end{pmatrix}$$

for *some* value of λ. X then, will be eigenvector of A with an eigenvalue λ. We can cast (5.24) as a homogeneous matrix equation

$$B \equiv (A - \lambda I)X = 0. \tag{5.25}$$

It can be shown that such an equation has at least one nontrivial solution for $X\,(\neq 0)$ whenever det $B = 0$.*

Consider now this equation

$$\det\,(A - \lambda I) = 0. \tag{5.26}$$

From our knowledge of the expansion of $n \times n$ determinant, it is easily seen that (5.27) is a algebraic equation in λ of nth degree. Now, from the fundamental theorem in algebra,** such an equation has at least one root, say λ_1. Then,

*See, for example E. Kreyszig, *Advanced Engineering Mathematics*, 4th ed., Section 7.10. Wiley, 1972

**See, example, Churchill & Brown, *Complex Variables*, McGraw-Hill, 2004

$$\det(A - \lambda_1 I) = 0$$

so that there exists an $X_1 \neq 0$ such that

$$(A - \lambda_1 I)X_1 = 0,$$

i.e., A has an eigenvector X_1 corresponding an eigenvalue λ_1, viz.

$$AX_1 = \lambda_1 X_1.$$

Note: If X_1 is an eigenvector, $\zeta X_1 (\zeta \neq 0)$ is also an eigenvector; but this will not be regarded as a distinct eigenvector as it is not an LI of X_1.

Corollary: A matrix A has at least as many distinct eigenvectors as there are distinct eigenvalues of A.

This can be seen as follows: Eq. (5.26) above has, in fact, n roots. But they may not all be distinct. There may be say, p distinct roots: $\lambda_1, \lambda_2, \ldots, \lambda_p$. The proof of the above theorem applies to each eigenvalue separately, implying that there are at least p eigenvectors $X_1, X_2, \ldots, X_p$ of A which are distinct (i.e., no X_1 is proportional to any $X_j, i, j = 1, 2, \ldots, p$). Of course, this itself does not guarantee that they are LI. In the following theorem, we shall prove that they indeed LI.

Theorem 5.8 Eigenvectors of a matrix corresponding to distinct eigenvalues are necessarily linearly independent.

Proof: Let A be a matrix with a set of p distinct eigenvalues $\lambda_1, \lambda_2, \ldots, \lambda_p$. Let $X_1, X_2, \ldots, X_p$ be a set of corresponding eigenvectors. If a given eigenvalue λ_i has more than one LI as eigenvectors, X_i represents *any* LC of these. First, consider the two eigenvalue equations corresponding to eigenvalues λ_1 and λ_2:

$$AX_1 = \lambda_1 X_1, \tag{5.27}$$

$$AX_2 = \lambda_2 X_2. \tag{5.28}$$

Now, were X_1 and X_2 linearly dependent, we would have $X_2 = \alpha X_1$ $(\alpha \neq 0)$. Then

$$AX_2 = \alpha AX_1 = \alpha\lambda_1 X_1 = \lambda_1 X_2. \tag{5.29}$$

As $X_2 \neq 0$, (5.28) and (5.29) imply $\lambda_1 = \lambda_2$, a contradiction since they are assumed to be distinct. Hence X_1 and X_2 are LI.

Now the proof proceeds by induction. Let the set $X_1, X_2, \ldots, X_q$ be linearly independent and let, if possible, the set $X_1, X_2, \ldots, X_q, X_{q+1}$ be linearly dependent. One would then have a relation of the form (with at least one $a_k \neq 0$)

$$\sum_{k=1}^{q+1} a_k X_k = 0. \tag{5.30}$$

In the above relation, a_{q+1} cannot vanish as otherwise it would imply that the set $\{X_1, X_2, \ldots, X_q\}$ is LD, which it is not by assumption. Dividing throughout by a_{q+1}, one has

$$X_{q+1} = -\sum_{k=1}^{q} \frac{a_k}{a_{q+1}} X_k \equiv -\sum_{k=1}^{q} \beta_k X_k, \tag{5.31}$$

where on the R.H.S. at least one term is nonzero since $X_{q+1} \neq 0$. Operating on (5.31) by A and using $AX_i = \lambda_i X_i$ in it,

$$AX_{q+1} = -\sum_{k=1}^{q} \beta_k AX_k,$$

$$\text{i.e.,} \quad \lambda_{q+1} X_{q+1} = -\sum_{k=1}^{q} \beta_k \lambda_k X_k,$$

$$\text{i.e.,} \quad -\lambda_{q+1} \sum_{k=1}^{q} \beta_k X_k = -\sum_{k=1}^{q} \beta_k \lambda_k X_k.$$

On account of LI of $\{X_1, X_2, \ldots, X_q\}$, one can compare coefficients of X_k on both sides to yield

$$\beta_k(\lambda_{q+1} - \lambda_k) = 0 \qquad k = 1, 2, \ldots, q.$$

As at least one β_k is nonzero; this implies

$$\lambda_{q+1} = \lambda_k \qquad \text{for some } k,\ 1, \leq k \leq q.$$

But this is contradiction as all λ's are assumed to be distinct. Hence if the set $\{X_1, X_2, \ldots, X_q\}$ is LI, so is the set $\{X_1, X_2, \ldots, X_{q+1}\}$. Now this result has already been proved for $q = 1$. Hence by induction it is proved for $q = p$. Hence, etc.

Corollary: An $n \times n$ matrix having n distinct eigenvalues have n LI eigenvectors. As any n LI vectors span vector space V of n-column vectors, this set can serve as a basis for V. Such a set is called a complete set. Thus, an $n \times n$ matrix having n distinct eigenvalues has a complete set of LI eigenvectors.

Comment: If a matrix A has a degenerate eigenvalue (one or more eigenvalue repeated more than once), then it may not generally have a complete set of n LI eigenvectors. We shall later on discuss the class of matrices which do necessarily have a complete set of n LI eigenvectors.

Examples

(i) The matrix $\begin{bmatrix} a & 1 \\ 0 & a \end{bmatrix}$ has a degenerate eigenvalue $\lambda_1 = a$, $\lambda_2 = a$. It, however, has only one LI eigenvector $\begin{bmatrix} 1 \\ 0 \end{bmatrix}$. (Verify !)

(ii) The matrix $\begin{bmatrix} a & 0 \\ 0 & a \end{bmatrix}$ has a degenerate eigenvalue $\lambda = a$. It, however, does have two distinct LI eigenvectors, say $\begin{bmatrix} 1 \\ 0 \end{bmatrix}$ and $\begin{bmatrix} 0 \\ 1 \end{bmatrix}$.

5.10.1 Characteristic (secular) equation

In general, determination of all the eigenvalues of an $n \times n$ matrix requires one to solve the nth degree equation

$$\det(A - \lambda I) = 0.$$

This equation is called the *characteristic equation.* However, certain simple combinations of all eigenvalues of A can be obtained much more directly. These results are proven in the theorem below. These combination of eigenvalues occur in physics quite often and often have direct physical significance.

Definition: For an $n \times n$ matrix A, $\mathrm{tr}A = \sum_{i=1}^{n} A_{ii}$ is called the trace of A.

Theorem 5.9

(i) $\mathrm{tr}A = \sum_{i=1}^{n} \lambda_i$ = sum of all eigenvalues of A.

(ii) $\det A = \prod_{i=1}^{n} \lambda_i$ = product of all eigenvalues of A.

Proof: Consider the eigenvalue equation $\det(A - \lambda I) = 0$, i.e., written explicity,

$$\begin{bmatrix} a_{11} - \lambda & a_{12} & \cdots & a_{nn} \\ a_{21} & a_{22} - \lambda & \cdots & a_{2n} \\ \vdots & \vdots & & \vdots \\ a_{n1} & a_{n2} & \cdots & a_{nn} - \lambda \end{bmatrix} = 0.$$

This, as we have already stated, is an algebraic equation of nth degree. Let us obtain some of the terms in the expansion of the determinant. Firstly, $O(\lambda^n)$ term is $(-\lambda)^n = (-)^n \lambda^n$. Now consider $O(\lambda^{n-1})$ term in the expansion. Expansion of the determinant contains a term $(a_{11} - \lambda)\,(a_{22} - \lambda)\ldots(a_{nn} - \lambda)$. This clearly has n $O(\lambda^{n-1})$ terms coming from a_{ii} in the ith bracket and $(-\lambda)$ in the remaining $(n - 1)$ brackets $(1 \leq i \leq n)$. These terms are $(-\lambda)^{n-1} \sum_{i=1}^{n} a_{ii}$.

Now, this is the only contribution to the $O(\lambda^{n-1})$ term because every other term in the expansion of the determinant contains at least two off-diagonal elements, none of which contain λ. Hence, these remaining terms in the expansion cannot contain $O(\lambda^{n-1})$ contribution.

Finally, $O(\lambda^0)$ term is obtained by setting $\lambda = 0$ in the L.H.S. and is simply det A. Hence the above characteristic equation reads,

$$(-\lambda)^n + (-\lambda)^{n-1}\sum_{i=1}^{n} a_{ii} + \ldots + det A = 0.$$

Now using the theorems on the roots of an nth degree algebraic equation,

$$\text{sum of all roots } = \sum_{i=1}^{n} \lambda_i = -\frac{\text{coefficient of } \lambda^{n-1}}{\text{coefficient of } \lambda^n} = \sum_{i=1}^{n} a_{ii} = \text{tr} A.$$

$$\text{Product of roots } = \text{product of all eigenvalues } \prod_{i=1}^{n} \lambda_i$$

$$= \frac{\text{coefficient of } \lambda^0}{(-)^n \text{ coefficient of } \lambda^n} = \det A.$$

Corollary: If det $A = 0$, at least one of the eigenvalues vanishes.

Example Consider the "Pauli Matrix" $\sigma_x := \begin{bmatrix} 0 & 1 \\ 1 & 0 \end{bmatrix}$. Let its eigenvalues be λ_1 and λ_2. From the above theorem,

$$\lambda_1 \lambda_2 = \det \sigma_x = -1,$$
$$\lambda_1 + \lambda_2 = \text{tr}\, \sigma_x = 0.$$

This, in fact, easily leads, in this simple 2×2 example, to the actual values of λ, viz $\lambda_1 = 1$, $\lambda_2 = -1$.

Theorem 5.10 For a real matrix, each eigenvalue is either real or if complex, its complex conjugate is also an eigenvalue.

Proof: Given a real matrix (all a_{ij} real), consider the characteristic equation det $(A - \lambda I) = 0$. This, as we saw, is an nth degree equation.

$$\sum_{m=0}^{n} b_m \lambda^m = 0,$$

where b_m are constructed from real linear combinations of a_{ij} (or of their products) and hence are all real. Hence for a real matrix A, the characateristic requation is an nth degree algebraic equation with real coefficients. In such a case, the roots (i.e., eigenvalues) are known to be either real or occuring in complex conjugate pairs.

Examples

(i) Consider the matrix $A = \begin{pmatrix} 1 & -2 \\ 1 & 1 \end{pmatrix}$. The eigenvalues are easily found by solving the characteristic equation. They are

$$\lambda = 1 \pm \sqrt{2}i,$$

i.e., they are a complex conjugate pair.

(ii) In the Example for the previous theorem, eigenvalues of σ_x are both real.

(iii) Consider the 3 × 3 matrix

$$0 = \begin{pmatrix} 1 & 0 & 0 \\ 0 & \frac{1}{\sqrt{2}} & \frac{1}{\sqrt{2}} \\ 0 & \frac{1}{\sqrt{2}} & \frac{1}{\sqrt{2}} \end{pmatrix}.$$

The eigenvalues are 1, $\frac{1+i}{\sqrt{2}}, \frac{1-i}{\sqrt{2}}$. Thus the first one is real, while the remaining are a complex conjugate pair.

We have seen in the previous theorem that even though they need not be all real, there is a restriction on the eigenvalues of a real matrix A. What can we say about the eigenvectors of A? Need they be all real? The answer is given in the following theorem.

Theorem 5.11

(i) The eigenvector of a real matrix A corresponding to a real eigenvalue λ can be chosen to be real.

(ii) The eigenvector X_1 of a real matrix A corresponding to a complex eigenvalue λ_1 is necessarily complex and its complex conjugate X_1^* is also an eigenvector of A corresponding to the eigenvalue λ_1^*.

Proof:

(i) Consider the eigenvalue equation

$$AX = \lambda X \tag{5.32}$$

where λ is real. Then either X is real (in which case part (i) of the theorem is proved) or writing $X = X_R + iX_I$ in terms of real and imaginary parts we have,

$$A(X_R + iX_I) = \lambda(X_R + iX_I). \tag{5.33}$$

Hence taking real and imaginary parts of the above equation

$$AX_R = \lambda X_R, \tag{5.34}$$

$$AX_I = \lambda X_I. \tag{5.35}$$

So, X_R and X_I (which are both real) are separately eigenvectors of A (not necessarily distinct, however) with the eigenvalue λ. Hence we could as well choose X_R and X_I as eigenvectors of A both of which are real. Hence, etc.

(ii) Now, consider the eigenvalue equation (λ_1 complex)

$$AX_1 = \lambda_1 X_1. \tag{5.36}$$

Suppose X_1 is real, then L.H.S. AX_1 would be real, while the R.H.S. $\lambda_1 X_1$ would be necessarily complex, leading to a contradiction. Hence X_1 must be complex.

Taking complex conjugate of the above equation ($A^* = A$ here),

$$AX_1^* = \lambda_1^* X_1^*,$$

proving the result (ii).

Now, we shall prove certain results about the eigenvalues and eigenvectors of real symmetric matrices as the latter are of common occurrence in Physics. We will comment on the physical applications of the results obtained at the end.

Theorem 5.12

(i) Eigenvalues of a real symmetric matrix are all real.

(ii) Eigenvectors of a real symmetric matrix can be chosen to be real (and it will further be shown that this can be chosen to be mutually orthogonal).

Proof:

(i) This result has already been proven in a different form. We have shown earlier that eigenvalues of an (abstract) hermitian operator are all real. Hermitian matrices are hermitian operators acting on the LVS of complex n-column vectors. Hence their eigenvalues are all real. A real symmetric matrix is a hermitian matrix and hence its eigenvalues are all real*.

(ii) We have seen that eigenvectors of a real matrix corresponding to a real eigenvalue can be chosen to be real. But all eigenvalues of a real symmetric matrix are real by virtue of (i), so that all its eigenvectors can be chosen to be real. Further, as this is a hermitian matrix, its eigenvectors can be chosen to be orthonormal as seen from comments following Theorem 5.18.

Comment: Real symmetric matrices occur very often in Physics. Examples are (i) the inertia tensor, (ii) polarizability tensor, (iii) metric tensor in general relativity. The eigenvalues and eigenvectors of these have physical significance. Eigenvalues of inertia tensor are the principle moments, which had better be real. Eigenvectors of inertia tensor are principle axes which

---

*Note that eigenvalues of a matrix are determined once the matrix is given. They have no need for a reference to any vector space on which it is supposed to act.

have directions in physical space and hence must also be represented by real column vectors. Similar statements can be made about polarizability tensor of a dielectric.

Definition: A matrix A is antihermitian if iA is hermitian, i.e., if $A = -A^\dagger$.

From previous theorems on hermitian matrices it follows that:

Theorem 5.13 Eigenvalues of an antihermitian matrix are purely imaginary and its eigenvectors can be chosen to be orthonormal.

Note: An antisymmetric real matrix is an antihermitian matrix to which the above theorem applies.

Now we turn to the results on the eigenvalues and eigenvectors of a real orthogonal matrix O. As noted earlier, in sec. 5.9, O is then a unitary matrix. Hence, its eigenvalues must be unimodular. Moreover, O is a real matrix. Hence an eigenvalue of O must be either real or they must occur in complex conjugate pairs. Hence eigenvalues of O must form a set

$$\underbrace{1, 1, \ldots, 1,}_{m} \ \underbrace{-1, -1, \ldots, -1,}_{q} \ \underbrace{e^{i\phi_1}, e^{-i\phi_1}, \ldots, e^{i\phi_p}, e^{-i\phi_p}}_{2p}$$

with $m + q + 2p = n$ being the dimension of O. This has applications, for example, in theory of rotations. There O is a 3×3 rotation matrix. Hence $m + q + 2p = 3$. Hence, $m + q \neq 0$ so that at least one eigenvalue is real and equal to $(\pm)1$. Moreover, for proper rotation, $\det O = 1$. Hence this eigenvalue must be $+1$. This proves that there exists at least one vector X such that

$$OX = X,$$

i.e., O leaves the direction of this vector unchanged. Moreover, unless $O = I$, the identity matrix, there is only one such LI vector (prove! Also see Problems 5.30 and 5.31). For a proper rotation matrix, this indeed is the rotation axis.

5.11 Transformation of Basis in a Vector Space

When we defined the matrix representation of a LO in Section 5.9.1, we chose a specific basis. It was remarked there that the matrix representation of the LO so obtained, is indeed dependent on the basis. We, now, wish to study how this matrix representation is affected when the basis is changed.

5.11.1 Transformation matrix

Firstly, we shall define the transformation matrix for the basis. We had chosen the basis $B = \{|e_1\rangle, |e_2\rangle, \ldots, |e_n\rangle\}$ to define the matrix representations. Now consider another basis $B' \equiv \{|e'_1\rangle, |e'_2\rangle, \ldots, |e'_n\rangle\}$. As B is a basis, every vector and therefore $|e'_i\rangle$ in particular can be expanded in terms of B:

$$|e'_i\rangle = \sum_{j=1}^{n} s_{ji}|e_j\rangle \qquad \text{for} \quad i = 1, 2, \ldots, n. \tag{5.37}$$

Conversely, we could as well expand any vector in B, say $|e_j\rangle$, in terms of the basis B':

$$|e_j\rangle = \sum_{k=1}^{n} t_{kj}|e'_k\rangle \qquad j = 1, 2, \ldots, n. \tag{5.38}$$

We may define matrices S and T by $s_{ji} = (S)_{ji}$ and $t_{kj} = (T)_{kj}$.

Now, substituting,

$$|e'_i\rangle = \sum_{k=1}^{n}\left[\sum_{j=1}^{n} t_{kj}s_{ji}\right]|e'_k\rangle = \sum_{k=1}^{n}(TS)_{ki}|e'_k\rangle.$$

In view of the LI of $|e'_k\rangle$, we may compare coefficients of $|e'_k\rangle$ and find

$$(TS)_{ki} = \delta_{ki}$$

or,

$$TS = I, \quad \text{i.e.,} \quad T = S^{-1}. \tag{5.39}$$

Thus, the transformation from one basis to another is given by a nonsingular matrix S, and the inverse matrix yields transformation from second basis to the first.

5.11.2 Transformation of matrix representations

(i) Now, consider a linear operator A. Consider its matrix representation defined in the basis B' of the LVS 'V'. Recalling the definition in Section 5.9.1,

$$A|e'_i\rangle \equiv \sum_{j} a'_{\ell i}|e'_\ell\rangle, \tag{5.40}$$

the corresponding matrix representation is given in terms of a matrix A' : $(\underline{A}')_{ji} = a'_{ji}$. We wish to relate A' to A. The L.H.S. of (5.40) is

$$\begin{aligned} A\sum_{j=1}^{n} s_{ji}|e_j\rangle &= \sum_{j} s_{ji}A|e_j\rangle \\ &= \sum_{j} s_{ji}\sum_{k} a_{kj}|e_k\rangle \\ &= \sum_{j} s_{ji}\sum_{k} a_{kj}\sum_{\ell} t_{\ell k}|e'_\ell\rangle \\ &= \sum_{\ell}(T\,\underline{A}\,S)_{\ell i}|e'_\ell\rangle, \end{aligned}$$

where use has been made of Eq. (5.22) in the second step and of Eq. (5.23) in the third step. We, thus, have $\underline{A}'$ obtained through

$$\sum_{\ell} a'_{\ell i}|e'_{\ell}\rangle = \sum_{\ell}(T\,\underline{A}\,S)_{\ell i}|e'_{\ell}\rangle.$$

As, $|e'_i\rangle$ are LI, we obtain

$$\underline{A}' = T\,\underline{A}\,S = S^{-1}\underline{A}S.$$

Matrices $\underline{A}$ and $\underline{A}'$ related by such a relation are said to be similar.

(ii) Next, we study transformation of vectors. Consider a vector $|x\rangle = \sum_i x^i|e_i\rangle$ expanded in basis $B = \sum x^{\ell\prime}|e'_{\ell}\rangle$ expanded in basis B'.

$$\text{From (5.38),} \quad |x\rangle = \sum_i x^i t_{\ell i}|e'_{\ell}\rangle.$$

Thus, $x^{\ell\prime} = \sum_i t_{\ell i}x^i$ or in terms of the matrix representation X and X' of $|x\rangle$ in B and B',

$$X' = TX = S^{-1}X.$$

5.11.3 Transformation between orthonormal bases

In Physics generally, and especially in Quantum Mechanics, one often makes a transformation of basis from one orthonormal basis to another. A familiar example is the rotation of cartesian coordinate frame in 3-D space chosen to describe the physical phenomena. Some other examples are the transformation from coordinate representation to momentum representation in Quantum Mechanics or that from the Schrodinger picture to the Heisenberg picture.

Theorem 5.14 An orthonormal basis is transformed into another orthonormal basis iff the matrix of transformation is unitary.

Proof:

(i) Consider two orthornormal bases $B \equiv \{|e_1\rangle, \ldots, |e_n\rangle\}$ and $B' \equiv \{|e'_i\rangle, \ldots, |e'_n\rangle\}$. They satisfy the orthonormality conditions:

$$\langle e_j|e_i\rangle = \delta_{ij} = \langle e'_j|e'_i\rangle. \tag{5.41}$$

Consider the transformation from B to B', viz,

$$|e'_i\rangle = \sum_{\ell} s_{\ell i}|e_{\ell}\rangle, \tag{5.42}$$

implying the dual vector equation

$$\langle e'_j| = \sum_k s^*_{kj} \langle e_k| \tag{5.43}$$

so that, using orthonormality condition on B', we have

$$\delta_{ij} = \langle e'_j|e'_i\rangle = \sum_k \sum_j s_{\ell i} s^*_{kj} \langle e_k|e_\ell\rangle \tag{5.44}$$

$$= \sum_k \sum_j s_{\ell i} s^*_{kj} \delta_{k\ell}$$

$$= \sum_k s^*_{kj} s_{ki} = \sum_k (S^\dagger)_{jk} (S)_{ki}$$

$$= (S^\dagger S)_{ji} \qquad i, j = 1, 2, \ldots, n. \tag{5.45}$$

Eq. (5.45) implies that $S^\dagger S = I$, i.e., the transformation matrix S is unitary.

(ii) Conversely, let B be an orthonormal basis and let S be a unitary transformation matrix from B to B'. Thus, $B' \equiv \{|e'_1\rangle, \ldots, |e'_i\rangle, \ldots, |e'_n\rangle\}$ with $|e'_i\rangle$ given by (5.42). We want to show that these $|e'_i\rangle$ satisfy the orthonormality condition (5.41)

$$\langle e'_j|e'_i\rangle = \delta_{ij}. \tag{5.46}$$

To this end, we calculate $\langle e'_j|e'_i\rangle$ using (5.42) and (5.43) exactly as done between equations (5.44) and (5.45) and find

$$\langle e'_j|e'_i\rangle = (S^\dagger S)_{ji},$$

but as S is given by to be unitary $(S^\dagger S)_{ji} = (I)_{ji} = \delta_{ij}$. Hence (5.46) is proved.

Unitary transformations and diagonalization of matrices play an important role in Quantum Mechanics. We shall soon discuss some important results regarding them. Before we proceed, however, we would like to study certain properties of unitary matrices.

Theorem 5.15 Column vectors of a unitary matrix form a complete orthornormal set, and conversely, a set of n mutually orthonormal n-column vectors stacked together as an $n \times n$ matrix constitute a unitary matrix.

Proof:

(i) Consider a unitary matrix U expressed in terms of its n columns $X_1, X_2, \ldots, X_n$, viz.,

$$U \equiv [X_1, X_2, \ldots, X_n]. \tag{5.47}$$

Then

$$U^\dagger = \begin{bmatrix} X_1^\dagger \\ X_2^\dagger \\ \vdots \\ X_n^\dagger \end{bmatrix} \tag{5.48}$$

Then we calculate $U^\dagger U$ and equate it to $I_{n\times n}$.

$$U^\dagger U = \begin{bmatrix} X_1^\dagger X_1 & X_1^\dagger X_2 & \cdots & X_1^\dagger X_n \\ X_2^\dagger X_1 & X_2^\dagger X_2 & \cdots & X_2^\dagger X_n \\ \vdots & \vdots & & \vdots \\ X_n^\dagger X_1 & X_n^\dagger X_2 & \cdots & X_n^\dagger X_n \end{bmatrix} = I = \begin{bmatrix} 1 & 0 & \cdots & 0 \\ 0 & 1 & \cdots & 0 \\ \vdots & \vdots & & \vdots \\ 0 & 0 & \cdots & 1 \end{bmatrix} \tag{5.49}$$

Comparing the elements of the two matrices one easily sees that

$$X_i^\dagger X_j = \delta_{ij}, \tag{5.50}$$

proving that the n columns of a unitary matrix form an orthonormal set. As they are mutually orthogonal, they also form an LI set of n vectors and hence serve as a basis for V_n. In other words, they are a complete orthonormal set.

(ii) Now we prove the converse. Let $X_1, X_2, \ldots, X_n$ be a set of n mutually orthonormal vectors, i.e., satisfy (5.50). Construct a matrix U by stacking together these as n columns of U, as in (5.47). Then calculate $U^\dagger U$ as in (5.49) and using $X_i^\dagger X_j = \delta_{ij}$ we see that it is equal to I. Hence U is unitary.

Corollary: The above theorem holds if the phrase "column vectors" is replaced by "row vectors". In the proof one uses $UU^\dagger = I$. (This is left as an exercise).

Comment 1 In part (ii) of the above theorem, one could consider any permutation of $X_1, X_2, \ldots, X_n$ and these will also form an orthonormal set. We would stack this permutation as successive columns of a (different) matrix U'. U' will also be a (different) unitary matrix. This is true for each permutation.

Comment 2 A real unitary matrix is orthogonal. Hence the theorem stated above also holds with "unitary matrix" replaced by "orthogonal matrix". Here, the column vectors under consideration are real and satisfy the orthonormality $X_i^T X_j = \delta_{ij}$.

5.12 Diagonalization of Matrices

Diagonalization of matrices occurs in a number of places in Physics. An operator on the vector space is represented by a matrix which, however, depends on the basis chosen to express it. It is often useful to know if there is a basis in which the matrix representation of the operator is diagonal. Apart from the obvious algebraic simplication that results when an operator is represented by a diagonal matrix, the importance of diagonalizing basis and diagonal representation results from the fact that this basis and the diagonal elements have generally a direct physical meaning. For example, consider the Hamiltonian operator for a physical system in Quantum Mechanics. As we shall see, for reasons explained later, its matrix representation can always be diagonalized. The diagonalizing basis consists of states which are called stationary states. States corresponding to definite energy levels and the diagonal elements turn out to be the corresponding energy eigenvalues (when the basis is orthonormal). In a similar manner, the diagonalizing basis of an inertia tensor gives the principle axes of the body, and the diagonal elements in this (orthonormal) basis give the principle moments, etc.

The above results are seen from the theorems that follow.

Consider a linear operator A on an n-dimensional vector space. Suppose it has a set of n LI eigenvectors $|e'_1\rangle, |e'_2\rangle, \ldots, |e'_n\rangle$. (Note that generally a given operator may not always have n LI eigenvectors.) Let the corresponding eigenvalues be $\lambda_1, \lambda_2, \ldots, \lambda_n$ (not necessarily distinct), i.e.,

$$A|e'_i\rangle = \lambda_i |e'_i\rangle. \tag{5.51}$$

Now, this set of eigenvectors constitute a basis B' for V and hence, can in particular, be chosen as a basis for a matrix representation of A. Call this matrix representation $\underline{A}'$, i.e.,

$$A|e'_i\rangle = \sum_j a'_{ji}|e'_j\rangle \qquad (\underline{A}')_{ji} = a'_{ji}. \tag{5.52}$$

Comparison of the R.H.S of (5.51) and (5.52) shows, using LI of the set $\{|e'_i\rangle\}$,

$$a'_{ji} = \lambda_i \delta_{ij}.$$

In other words,

$$\underline{A}' = \begin{bmatrix} \lambda_1 & 0 & \cdots & 0 \\ 0 & \lambda_2 & \cdots & 0 \\ \vdots & \vdots & & \vdots \\ 0 & 0 & \cdots & \lambda_n \end{bmatrix} \equiv \text{diag}\,(\lambda_1, \lambda_2, \ldots, \lambda_n). \tag{5.53}$$

We thus see that if an operator has a complete set of eigenvectors, there necessarily exists a basis (that of this set), in which it is represented by a diagonal matrix.

Now, let us proceed further. Consider the matrix representation of $\underline{A}$ in any arbitrary basis $B \equiv \{|e_1\rangle, |e_2\rangle, \ldots, |e_n\rangle\}$, viz.,

$$A|e_i\rangle = \sum a_{ji}|e_j\rangle \qquad (\underline{A})_{ji} \equiv a_{ji}. \tag{5.54}$$

Let this basis be related to the basis B' through a matrix $\underset{\sim}{S}$:

$$|e_i'\rangle \equiv \sum_j s_{ji}|e_j\rangle$$

Then, as we have already established in § 5.11.2

$$\underline{A}' = S^{-1}\underline{A}S \tag{5.55}$$

but $\underline{A}'$ is diagonal. Thus, the matrix $\underline{A}$ can be cast in a diagonal form $\underline{A}'$ by a similarity transformation S provided A has n linearly independent eigenvectors (these being the column-vector representations of $|e_1'\rangle, \ldots, |e_n'\rangle$). We now sutdy the structure of this diagonalizing matrix S.

Theorem 5.16 If a matrix $\underline{A}$ has complete set of eigenvectors, it can be diagonalized by a similarity transformation. The matrix S of this similarity transformation has theses n eigenvectors as its columns.

Proof: Existence of a similarity transformation that diagonalizes $\underline{A}$ has already been proven. Consider, now, the eigenvalue equation

$$A|e_i'\rangle = \lambda_i|e_i'\rangle. \tag{5.56}$$

We substitute for $|e_i'\rangle$ from Eq. (5.42) to find

$$A\sum_j s_{ji}|e_j\rangle = \lambda_i \sum_j s_{ji}|e_j\rangle. \tag{5.57}$$

We use Eq. (5.54) in the above to obtain,

$$\sum_j s_{ji}\sum_k a_{kj}|e_k\rangle = \lambda_i \sum_k s_{ki}|e_k\rangle. \tag{5.58}$$

On equating coefficients of the LI vectors $|e_k\rangle$, we obtain

$$a_{kj}s_{ji} = \lambda_i s_{ki}. \tag{5.59}$$

We now construct the ith column vector $S_{(i)}$ of S.

$$S_{(i)} = \begin{bmatrix} s_{1i} \\ s_{2i} \\ \cdots \\ s_{ni} \end{bmatrix}.$$

The above equation, can then be written as

$$\underline{A}S_{(i)} = \lambda_i S_{(i)}, \tag{5.60}$$

proving that the columns of $S_{(i)}$ are indeed eigenvectors of $\underline{A}$.

The converse of the above is also true.

Converse of Theorem 5.16 : If a matrix A can be diagonalized by a similarity transformation S, then n columns vectors of S are eigenvectors of $\underline{A}$. Moreover, as S is a nonsingular matrix, the n column vectors are LI, and thus form a complete set* of eigenvectors of A.

Thus, given a matrix A with a complete set of eigenvectors $X_1, \ldots, X_n$, one diagonalizing matrix S can be obtained (leading to $S^{-1}AS$ diagonal) by putting together this complete set of eigenvectors of S as column vectors of S, i.e.,

$$S = (X_1, X_2, \ldots, X_n).$$

Note however that this diagonalizing matrix S is not unique because

(i) the order of eigenvectors in the set can be varied;

(ii) the (nonzero) normalization of each of X's can be varied independently;

(iii) in the case of one or more degenerate eigenvalues, the set itself can be chosen in infinitely many ways (see Problem 5.61). This is in addition to the arbitrariness mentioned in (i) and (ii) above.

Further, instead of evaluating $S^{-1}AS$, if one replaces S by another diagonalizing matrix S' in which the order of eigenvectors is reshuffled, $S'^{-1}AS'$ differs from $S^{-1}AS$ (though both are diagonal) in that the order of diagonal elements in $S'^{-1}AS'$ is correspondingly altered as compared to that in $S^{-1}AS$ (see Problem 5.61).

However, if S' differs from S by the variations listed above in (ii) and/or (iii), $S'^{-1}AS' = S^{-1}AS$ (see problem 5.61).

An important corollary to this theorem and the preceeding one on the properties of unitary matrices (Theorem 5.15) is the following.

Theorem 5.17 A matrix can be diagonalized by a unitary transformation iff it has the complete set of mutually orthogonal eigenvectors.

For if A has n mutually orthogonal eigenvectors, they can be chosen as the *orthonormal* set, say $(X_1, \ldots, X_n)$. Then the diagonalizing matrix $S \equiv (X_1, \ldots, X_n)$ is unitary by Theorem 5.15. Conversely, if the diagonalizing matrix for A viz. $U \equiv (Y_1, \ldots, Y_n)$ is unitary, A has $Y_1, Y_2, \ldots, Y_n$ as eigenvectors and these are mutually orthogonal.

A further corollary is: a matrix can be diagonalized by a real orthogonal transformation iff it has a complete set of mutually orthonormal real eigenvectors.

5.13 Main Theorem on Diagonalization

Theorem 5.18 The necessary and sufficient condition that a matrix A can be diagonalized by a unitary transformation is that A is normal, i.e., $[A, A^\dagger] = 0$.

*See, for example, E. Kreyszig, *Advanced Engineering Mathematics*, Chapter 7, Wiley, 1972

Proof:

(1) Let A be normal. We have already proven the following properties of a normal matrix; viz,

(a) eigenvectors of A are orthogonal;
(b) If $AX = \lambda X$, then $A^{\dagger}X = \lambda^* X$.

We shall utilize these properties and prove that A can be diagonalized by a unitary transformation. In this, we proceed by induction.

Let A be an $r \times r$ matrix. Let assume that the theorem is true for $r = n$. The theorem is evidently true for $r = 1$. Now consider $r = n + 1$, i.e., consider A to be an $(n + 1) \times (n + 1)$ matrix. A has $n + 1$ eigenvalues. Let λ_1 be any one of these, and let X_1 be the corresponding normalized eigenvector. $AX_1 = \lambda_1 X_1$; $X_1^{\dagger} X_1 = 1$. Now, in the space of $(n + 1)$ dimensional column vectors, one can always construct an orthonormal basis B, in which X_1 is the first vector: $B = (X_1, Y_1, \ldots, Y_n)$ (one can use the Schmidt procedure). Now construct the $(n + 1) \times (n + 1)$ matrix $V \equiv (X_1, Y_1, \ldots, Y_n)$ whose columns are these $n + 1$ basis vectors. By Theorem 5.15, V is unitary. Now consider

$$V^{\dagger}AV = \begin{bmatrix} X_1^{\dagger} \\ Y_1^{\dagger} \\ \vdots \\ Y_n^{\dagger} \end{bmatrix} A(X_1, Y_1, \ldots, Y_n).$$

This has the following form:

$$= \begin{bmatrix} X_1^{\dagger}AX_1 & X_1^{\dagger}AY_1 & \cdots & X_1^{\dagger}AY_n \\ Y_1^{\dagger}AX_1 & & & \\ \vdots & & B_{n\times n} & \\ Y_n^{\dagger}AX_1 & & & \end{bmatrix}$$

where $B_{n\times n}$ is an $n \times n$ matrix.
Now,

$$X_1^{\dagger}AX_1 = \lambda_1 X_1^{\dagger}X_1 = \lambda_1; \text{ and}$$

$$\left.\begin{aligned} &Y_i^{\dagger}AX_1 = \lambda_1 Y_i^{\dagger}X_1 = 0 \\ &X_1^{\dagger}AY_i = (Y_i^{\dagger}A^{\dagger}X_1)^{\dagger} = (\lambda_1^* Y_i^{\dagger}X_1)^* = 0 \end{aligned}\right\} i = 1, 2, \ldots, n$$

owing to the orthogonality of X_1 to Y_i's. In the last equation we have used property (b) listed at the beginning of proof. Thus

$$V^{\dagger}AV = \left[\begin{array}{c|ccc} \lambda_1 & 0 & \cdots & 0 \\ \hline 0 & & & \\ \vdots & & B & \\ 0 & & & \end{array}\right].$$

What we have done till now is to find a unitary transformation V that transforms A in to a block diagonal form in which there is an $n \times n$ block matrix B. The purpose of doing this to make use of our assumption that the theorem under consideration holds for an $n \times n$ normal matrix. However, in order that this can be used in an inductive procedure, we must establish that B is a normal matrix. This must follow from normal nature of A. To see this, we calculate

$$(V^{\dagger}AV)^{\dagger}(V^{\dagger}AV) = \left[\begin{array}{c|ccc} |\lambda_1|^2 & 0 & \cdots & 0 \\ \hline 0 & & & \\ \vdots & & B^{\dagger}B & \\ 0 & & & \end{array}\right].$$

In a similar manner, we find

$$(V^{\dagger}AV)(V^{\dagger}AV)^{\dagger} = \left[\begin{array}{c|ccc} |\lambda_1|^2 & 0 & \cdots & 0 \\ \hline 0 & & & \\ \vdots & & BB^{\dagger} & \\ 0 & & & \end{array}\right].$$

But

$$(V^{\dagger}AV)^{\dagger}(V^{\dagger}AV) = V^{\dagger}A^{\dagger}VV^{\dagger}AV = V^{\dagger}A^{\dagger}AV \tag{5.61a}$$

and also

$$(V^{\dagger}AV)(V^{\dagger}AV)^{\dagger} = V^{\dagger}AA^{\dagger}V, \tag{5.61b}$$

using the unitary nature of $V : VV^{\dagger} = I$. Now, as $A^{\dagger}A = AA^{\dagger}$, the two matrices on the right hand sides of (5.61) and (5.62) must be equal. This leads us to

$$B^{\dagger}B = BB^{\dagger},$$

i.e., that B is an $n \times n$ normal matrix. Thus, by our assumption, there exists an $n \times n$ unitary matrix W_1 such that $W_1^\dagger B W_1$ is a diagonal matrix. In fact, we can determine diagonal element of $W_1^\dagger B W_1$ in terms of the eigenvalues of A. We know that a similarity transformation does not alter the eigenvalues. Hence $V^\dagger A V$ has the same set of eigenvalues as A. Because of its given block diagonal nature $V^\dagger A V$ has one eigenvalue λ_1. Hence eigenvalues of B must be the remaining n eigenvalues of A, say $\lambda_2, \ldots, \lambda_{n+1}$. Thus, there exists W_1 such that

$$W_1^\dagger B W_1 = \text{diag}\,(\lambda_2, \lambda_3, \ldots, \lambda_{n+1}).$$

Now, out of this $n \times n$ unitary transformation W_1, we shall construct a $(n+1) \times (n+1)$ unitary transformation. Consider,

$$W = \begin{bmatrix} 1 & 0 \\ 0 & W_1 \end{bmatrix}.$$

Evidently,

$$\begin{aligned} W W^\dagger &= \begin{bmatrix} 1 & 0 \\ 0 & W_1 \end{bmatrix} \begin{bmatrix} 1 & 0 \\ 0 & W_1^\dagger \end{bmatrix} = \begin{bmatrix} 1 & 0 \\ 0 & W_1 W_1^\dagger \end{bmatrix} = \begin{bmatrix} 1 & 0 \\ 0 & I_{n \times n} \end{bmatrix} \\ &= I_{n+1 \times n+1} \end{aligned}$$

showing that W is unitary. We now apply this unitary transformation to $V^\dagger A V$. Thus we compute

$$\begin{aligned} W^\dagger (V^\dagger A V) W &= \begin{bmatrix} 1 & 0 \\ 0 & W_1^\dagger \end{bmatrix} \begin{bmatrix} \lambda_1 & 0 \\ 0 & B \end{bmatrix} \begin{bmatrix} 1 & 0 \\ 0 & W_1 \end{bmatrix} \\ &= \begin{bmatrix} \lambda_1 & 0 \\ 0 & W_1^\dagger B W_1 \end{bmatrix} \\ &= \begin{bmatrix} \lambda_1 & 0 \\ 0 & \begin{matrix} \lambda_2 & & \\ & \ddots & \\ & & \lambda_{n+1} \end{matrix} \end{bmatrix} \end{aligned}$$

We can write this as

$$(V W)^\dagger A (V W) = \text{diag}\,(\lambda_1, \lambda_2, \ldots \lambda_{n+1}).$$

Further, $V W$ is unitary since each of V and W is unitary, so

$$(V W)(V W)^\dagger = V W W^\dagger V^\dagger = V V^\dagger = I.$$

We have thus found an $(n+1) \times (n+1)$ unitary matrix transformation $U \equiv VW$ which diagonalizes A:

$$U^{\dagger}AU = \text{diag}\,(\lambda_1, \dots, \lambda_{n+1})$$

thus proving the first part of the theorem by induction.

(2) Now we prove the converse. Let for a square matrix A there exist a unitary transformation U such that

$$U^{\dagger}AU = D = \text{a diagonal matrix.}$$

Using $DD^{\dagger} = D^{\dagger}D$ for a diagonal matrix we now compute

$$(U^{\dagger}AU)^{\dagger}(U^{\dagger}AU) = D^{\dagger}D = DD^{\dagger} = (U^{\dagger}AU)(U^{\dagger}AU)^{\dagger}$$

Evaluating the two expressions as in Eq. (5.61)

$$U^{\dagger}A^{\dagger}AU = U^{\dagger}AA^{\dagger}U$$

As U is an invertible matrix (or equivalently premultiplying by U and postmultiplying by $U^{\dagger}$) we find

$$A^{\dagger}A = AA^{\dagger}$$

i.e., A is normal.

Corollary: A normal matrix always has a complete set of orthonormal vectors. This follows from the corollary to Theorem 5.17, as A can be diagonalized by a unitary transformation. Note that this is irrespective of whether A has degenerate eigenvalues. In other words, while a general $n \times n$ matrix with degenerate eigenvalues need not necessarily have a complete set of eigenvectors, a normal matrix necessarily does.

Problems

5.1 Show that if A and B are linear operators on V, then so are $A + B$ and AB.

5.2 Show that if the "matrix elements" of A viz. $\langle b|A|a\rangle$ for all $|a\rangle$ and $|b\rangle \in V$ are known, then $A|a\rangle$ is known for arbitrary $|a\rangle \in V$ and conversely.

5.3 Consider the 3D real vector space. A linear operator 'A' is completely defined by

$$A\hat{\mathbf{i}} = \hat{\mathbf{j}} \quad A\hat{\mathbf{j}} = -\hat{\mathbf{j}} \quad A\hat{\mathbf{k}} = \hat{\mathbf{i}} + \hat{\mathbf{k}}.$$

(i) Find the domain and the range of A.

(ii) Find A^2 by obtaining its action on the basis vectors.

(iii) Find the vectors $\mathbf{v}$ such that $A\mathbf{v} = 0$
Now consider further an operator B defined by

$$B\hat{\mathbf{i}} = \hat{\mathbf{k}} \quad B\hat{\mathbf{j}} = \hat{\mathbf{i}} \quad B\hat{\mathbf{k}} = \hat{\mathbf{j}}.$$

(iv) Find B^2 and show that $B^3 = I$.
(v) Find $A + \frac{B}{2}$ and BA.
(vi) Are there solutions to $B\mathbf{v} = 0$ (other than $\mathbf{v} = 0$)?

5.4 Show that for an arbitrary linear operator 'A' on a vector space V, the set of all vectors $\{A|a\rangle, \ |a\rangle \in V\}$ is a vector subspace V_1. Further, suppose that V is n-dimensional and $A|b\rangle = |0\rangle$ has m L I solutions, then V_1 has dimension $n - m$.

5.5 Consider a linear operator 'A' on a vector space V. Suppose $A|b\rangle = 0$ has no solution except $|b\rangle = |0\rangle$. Show that every vector $|a\rangle \in V$ can uniquely be represented as $A|c\rangle$ where $|c\rangle \in V$. Show further, that left inverse of A exists in this case and is given by $A_\ell^{-1}|a\rangle = |c\rangle$.

5.6 Consider the set of real infinite uniformly convergent series in x. Show that this set forms a real linear vector space $V : f(x) \in V \to f(x) = \sum_{n=0}^{\infty} a_n x^n$, ($a_n$ real). Consider the operator $A = \frac{d}{dx}$. Show that A maps V onto V. (i.e. for every $f(x) \in V$, $Af(x) \in V$ and every $g(x) \in V$ is expressible as $Af(x)$ for some $f(x) \in V$).

(i) Does A have a left inverse? [Hint : Consider $f(x) =$ constant.]
(ii) Does A have a right inverse? $\left[\text{Hint : Consider } B \text{ defined by } Bf(x) = \sum_{n=0}^{\infty} a_n \frac{x^{n+1}}{n+1}\right]$. What is the meaning of operation B?
(iii) Is the right inverse unique?
(iv) Is there an isomorphism between this vector space and the one considered in the example in Section 5.2?

5.7 Show that for $P = |e\rangle\langle e|$, $P^2 = P$ if $\langle e|e\rangle = 1$. Show further that $\{P|b\rangle; b \in V\}$ constitutes a subspace of V. what is its dimensionality?

5.8 Let $|e_1\rangle, \ |e_2\rangle, \ldots, |e_n\rangle$ constitute an orthonormal basis in V. Then show that $\sum_i |e_i\rangle\langle e_i| = I$. This relation is called the completeness relation.

5.9 Show that $P_j \equiv |e_j\rangle\langle e_j|$ [where $\langle e_i|e_j\rangle = \delta_{ij}$] satisfy $P_i P_j = P_i \delta_{ij}$ viz. the condition for orthogonal projection operators.

5.10 Show that the projection operator $P_j = |e_j\rangle\langle e_j|$ defined in Problem 5.8 is hermitian. Is it also unitary? Consider the cases where dimension of the space $n = 1, 2, \ldots$.

5.11 Can an operator be both hermitian and unitary? Give an example of a 2×2 matrix (operator) to justify your conclusion.

5.12 An operator in an n-dimensional vector space is defined by $A = \sum_{i=1}^{n} \lambda_i |e_i\rangle\langle e_i|$ where $\{|e_i\rangle\}$ are an orthonormal basis. Show that (i) A is hermitian iff λ_i are all real (ii) A is unitary iff $|\lambda_i| = 1$ for $i = 1, 2, \ldots, n$. Hence show that this operator A is simultaneously hermitian and unitary only when $A = \sum_{i=1}^{n} \pm|e_i\rangle\langle e_i|$.

5.13 Show that every operator A can be written as a linear combination of a hermitian operator and an antihermitian operator in a unique manner.

5.14 Let A be an operator defined by

$$A\hat{\mathbf{i}} = \hat{\mathbf{j}} \qquad A\hat{\mathbf{j}} = -\hat{\mathbf{j}} \qquad A\hat{\mathbf{k}} = \hat{\mathbf{i}} + \hat{\mathbf{k}}.$$

Find $A^\dagger$ and express its action on unit basis vectors.

5.15 Let $|e_1\rangle, |e_2\rangle, \ldots |e_n\rangle \ldots$ be an orthonormal basis and let $|e'_1\rangle, |e'_2\rangle, \ldots, |e'_n\rangle \ldots$ be another orthonormal basis. Define U by

$$|e'_i\rangle = U|e_i\rangle.$$

Show that U is unitary. Prove the converse.

5.16 Let H be a hermitian operator. Define an operator

$$\exp(iH) \equiv \sum_{n=0}^{\infty} \frac{(iH)^n}{n!}.$$

Find the adjoint of this operator and show that this operator is unitary.

5.17 Show that a projection operator P has an inverse iff $P = I$. What is the meaning of this statement?

5.18 Consider the projection operator P defined on a real 3-D vector space:

$$P_1\hat{\mathbf{i}} = \frac{\hat{\mathbf{i}} + \hat{\mathbf{j}}}{2}; \quad P_1\hat{\mathbf{j}} = \frac{\hat{\mathbf{i}} + \hat{\mathbf{j}}}{2}; \quad P_1\hat{\mathbf{k}} = 0$$

which projects V onto a subspace V_1. Verify that P_1 is a projection operator. Obtain the projection operator P_2 that projects onto the orthogonal subspace V_2 such that $V_1 \oplus V_2 = V$.

5.19 Find the eigenvalues of $\sigma_1 = \begin{pmatrix} 0 & 1 \\ 1 & 0 \end{pmatrix}$

(i) by direct calculation,
(ii) by finding sum and product of eigenvalues.

Also find the two normalized eigenvectors.

5.20 Find eigenvalues and eigenvectors of $\begin{pmatrix} 1 & 2 \\ 0 & 1 \end{pmatrix}$. How many eigenvectors does it have? Are all eigenvalues real? Is the matrix hermitian? Comment.

5.21 Find the eigenvalues and eigenvectors of

$$\begin{bmatrix} 1 & 1 & 0 \\ 1 & 0 & 1 \\ 0 & 1 & 1 \end{bmatrix}$$

5.22 Show that the matrix $\begin{bmatrix} a & b \\ 0 & c \end{bmatrix}$ has two LI eigenvectors if $a \neq c$ and only one LI eigenvector if $a = c$ (assume $b \neq 0$). Study the limit $c \to a$.

5.23 For a triangular matrix $a_{ij} = 0;\ i < j$ (or $i > j$), show that the diagonal elements of a triangular matrix are its eigenvalues.

5.24 Show that A and A^T have the same eigenvalues. Do they (necessarily) have the same eigenvectors?

5.25 Show that if $|a\rangle$ is an eigenvector of A with eigenvalue λ, then it is also an eigenvector of A^m with eigenvalue λ^m (m is a positive integer). Construct a simple example to show that if A^m has $|b\rangle$ as an eigenvector, it need not be an eigenvector of A at all.

5.26 Show that if A has a zero eigenvalue, left inverse A_ℓ^{-1} does not exist. If B does however have a left inverse, every eigenvector of B is an eigenvector of B_ℓ^{-1} with reciprocal eigenvalue.

5.27 Show that if two operators A and B have a complete set of common eigenvectors, then they commute. (Here, the set of common eigenvectors span the LVS.)

5.28 Show that if two operators A and B commute every eigenvector of A corresponding to a nondegenerate eigenvalue is also an eigenvector of B and vice versa.

5.29 Even when two operators do not commute, they can have common eigenvector(s), but the common eigenvectors cannot form a complete set. Verify this statement with the help of the two matrix operators A and B given below [see Problem 5.20].

$$A = \begin{bmatrix} a & b \\ 0 & c \end{bmatrix}; \quad B = \begin{bmatrix} a & c \\ 0 & f \end{bmatrix} \qquad \begin{array}{l} a \neq c,\ a \neq f \\ ab + cd \neq ad + bf \end{array}$$

(In the hydrogen atom problem, L_x, L_y and L_z have ψ_{100} (the ground state wavefunction) as a common eigenvector with eigenvalues all zero. Yet L_x, L_y, L_z do not commute.)

5.30 Pure rotation is represented by a 3×3 orthogonal matrix O with determinant one. Show that its eigenvalues must be 1, $e^{i\phi}$, $e^{-i\phi}$. Show further that for a nontrivial rotation, $\phi \neq 2n\pi$ and that $-1 \leq \text{tr}O < 3$.

5.31 A rotation in $(2n+1)$-dimensional Euclidean space is represented by a $(2n+1) \times (2n+1)$ orthogonal matrix. Show that any pure rotation in this space leaves vectors along at least one direction unchanged. What would the corresponding result be for pure rotations in $2n$ dimensions?

5.32 Show that if a matrix is both hermitian and unitary, then its eigenvalues are all ± 1. Verify that the following matrix is both hermitian and unitary and write down its eigenvalues by inspection.

$$A = \begin{bmatrix} 0 & 1 & 0 & 0 \\ 1 & 0 & 0 & 0 \\ 0 & 0 & 0 & 1 \\ 0 & 0 & 1 & 0 \end{bmatrix}$$

[Hint: What is tr A?]

5.33 Show that the eigenvalues of $A^\dagger A$ for any $n \times m$ matrix A, are necessarily nonnegative.

5.34 Show that the only possible eigenvalues of a projection operator $P(P^2 = P)$ are 0 and 1.

5.35 A linear operator 'A' is defined by

$$A\mathbf{i} = \mathbf{j} \quad A\mathbf{j} = \mathbf{k} \quad A\mathbf{k} = \mathbf{i}.$$

Find the matrix representation of A. What are the eigenvalues and eigenvectors of A? Is A orthogonal? Can you predict these eigenvalues by inspection of A? What is the matrix representation of A in the basis

$$\mathbf{i}' = \frac{\mathbf{i}+\mathbf{j}}{\sqrt{2}}, \quad \mathbf{j}', = \frac{\mathbf{i}-\mathbf{j}}{\sqrt{2}}, \quad \mathbf{k}' = \mathbf{k}.$$

What is the transformation matrix S that takes one from the first cartesian coordinate frame to the primed one?

5.36 Show that the trace and the determinant of a square matrix remain unchanged under a similarity transformations. Show how the invariance of eigenvalues under similarity transformation implies this result.

5.37 Show that if all the eigenvalues of a matrix are equal and its eigenvectors form a complete set, then it is a multiple of the unit matrix. Give a simple counter-example to show that the completeness of eigenvectors in the statement of the problem is indeed necessary.

5.38 Diagonalize the following matrix

$$A = \begin{bmatrix} 1 & 0 & 1 \\ 0 & 1 & 0 \\ 1 & 0 & 1 \end{bmatrix}.$$

Is diagonalizing matrix S unique? Can it be chosen to be orthogonal? Must it be orthogonal? Must it be real?

5.39 Show that the hermitian, normal or the unitary nature of a matrix is unchanged under a unitary transformation.

5.40 Show that a real symmetric matrix goes into another real symmetric matrix under a real orthogonal transformation.

5.41 Show that if the eigenvalues of a matrix A are real and its eigenvectors (can be chosen to) form an orthonormal complete set, then A is hermitian. Show by a simple counter example that the second condition in this problem is indeed a necessary one.

5.42 Show that for any (rectangular / square) A, $AA^\dagger$ and $A^\dagger A$ can always be diagonalized by a unitary transformation.

5.43 Show that if A is normal and triangular, then A is diagonal.

5.44 Show that a normal matrix is (a) hermitian iff all its eigenvalues are real (ii) unitary iff all its eigenvalues are unimodular.

5.45 Two square matrices U and H are related by $U = e^{iH}$. Show (assuming H can be diagonalized) that (i) U is unitary iff H is hermitian (ii) det $U = \exp\{i \text{ tr } H\}$.

5.46 Show that every square matrix A can be expressed as $A = HU$ where H is hermitian and U is unitary. (Hint: Use the result of Problem 5.42 and 5.33.) This is the matrix analogue to $z = Re^{i\phi}$ for a complex number where R is real and $e^{i\phi}$ is unimodular.

5.47 Show that a normal matrix A with eigenvalues λ_n and orthonormal complete set of eigenvectors $X_n (n = 1, 2, \ldots, m)$ can be expressed as

$$A = \sum_n \lambda_n X_n X_n^\dagger.$$

Verify this for $A = \begin{bmatrix} 0 & 1 \\ 1 & 0 \end{bmatrix}$ by constructing the right hand side.

5.48 Two hermitian matrices have the same set of eigenvalues (with the same multiplicities). Show that they must be related to each other by a unitary transformation. Verify this result for $\sigma_1 = \begin{pmatrix} 0 & 1 \\ 1 & 0 \end{pmatrix}$ and $\sigma_3 = \begin{pmatrix} 1 & 0 \\ 0 & -1 \end{pmatrix}$.

5.49 Let $\lambda_{\min}$ and $\lambda_{\max}$ be the smallest and largest eigenvalues of a hermitian matrix H. Show that for any $X \neq 0$, $X^\dagger HX$ is real and that

$$\lambda_{\min} \leq \frac{X^\dagger HX}{X^\dagger X} \leq \lambda_{\max}.$$

Hence show that $X^\dagger HX/X^\dagger X$ attains its extremum over $X(\neq 0)$ only when X is an eigenvector of H.

5.50 Let $X_{(i)} (i = 1, 2, \ldots, m)$ be a set of n vectors. Define an $m \times m$ matrix A by its matrix elements

$$A_{ij} = X_{(i)}^\dagger X_{(j)}.$$

Show that A is a hermitian matrix with nonnegative eigenvalues. (Hint: If $Y = \begin{bmatrix} y_1 \\ \vdots \\ y_m \end{bmatrix}$ be an eigenvector, $\lambda = Y^\dagger AY/Y^\dagger Y$. Show this to be nonnegative.)

5.51 A and B are both $n \times n$ matrices with complete sets of eigenvectors. Show that a complete set of common eigenvectors can be found iff $[A, B] = 0$. Hence show that A and B can be simultaneously diagonalized (i.e. diagonalized by the same similarity transformation) iff $[A, B] = 0$.

On § 5.3

5.52 Verify that the association $a|x\rangle + \beta|y\rangle$ in $V \to \alpha^*\langle x| + \beta^*\langle y|$ in V' is *uniquely* defined. Do this by choosing a basis in V. Further verify that V' forms a LVS.

5.53 Show that if V' is the vector space dual to V, then V is dual to V'.

5.54 Verify that $\langle z|x\rangle$ is linear in both $|x\rangle$ and $\langle z|$.

Miscellaneous

5.55 Show that $(A + B)^\dagger = A^\dagger + B^\dagger$; $(AB)^\dagger = B^\dagger A^\dagger$.

5.56 Show that every operator can be decomposed uniquely into a hermitian and an antihermitian part.

5.57 For an identity operator I and any LOA, show that $AI = IA = A$.

5.58 Prove the observations made at the end of Section 5.6.

5.59 Prove the observations regarding non-uniqueness of S following Theorem 5.16.

5.60 Show that if an operator has every (non-null) vector in V as its eigenvector, then it is a multiple of identity.

5.61 Consider the triangular matrix

$$A = \begin{bmatrix} 1 & 1 & 0 \\ 1 & 1 & 0 \\ 0 & 0 & 2 \end{bmatrix}.$$

(a) Write down the eigenvalues by inspection. Comment on the possible degeneracy.

(b) Does A have a complete set of eigenvectors?

(c) Construct at least four distinct diagonalizing matrices using the three ambiguities in such a construction mentioned following Theorem 5.16.

(d) Calculate $S^{-1}AS$ for the four cases and compare.

6 Complex Variables – I

In this and the next two chapters we shall discuss basics of complex variables and some of their applications. Familiarity with the elements of complex variables will be assumed. In this chapter, we shall discuss differentiability of a function of the complex variable z and then introduce the concept of analytic functions. We shall then discuss a number of simple analytic functions.

In the next chapter we shall introduce contour integrals and a number of results based on the Cauchy integral formula. In particular, we shall introduce the Taylor series and Laurent series. An important result from the point of view of practical calculations, viz. the Residue Theorem, will also be established.

The last chapter in this book will contain a number of distinct topics useful in Physics, viz. (a) analytic continuation, (b) conformal mapping, (c) method of steepest descent (d) Riemann surfaces.

From the point of view of Physics, the results outlined above will be of great utility and necessity. We name a few of the applications:

(a) In Quantum Mechanics, use of complex numbers is a necessity. The quantum mechanical wavefunction is (generally) a complex number. The imaginary unit $i(=\sqrt{-1})$ occurs in the basic Schrödinger equation itself.

(b) Certain problems in 2-D electrostatics are rather easily solved with the help of conformal mapping. More on this in Chapter 8.

(c) Contour integrals are useful in a variety of situations such as:

   i) Green's functions for a wave equation can be conveniently represented by a contour integral.
   ii) The method of contour integrals is useful as a purely mathematical device to evaluate certain types of definite integrals.
   iii) A number of special functions used in Physics find convenient contour integral representations (see second volume of this book dealing with advanced topics.)

(d) From a purely mathematical point of view, the treatment of ordinary differential equations (see the accompanying volume *Mathematical Physics: Advanced Topics*) becomes greatly simplified if the independent variable is chosen as complex z rather than real x.

(e) The theory of analytic functions leads to many powerful and beautiful results and these have been employed in advanced Physics in a variety of situations.

6.1 Introduction

Complex variables involve the imaginary unit 'i' satisfying $i^2 = -1$. A complex variable z depends on two real variables x and y, and can be written as $z = x + iy$. Alternatively, it can also be thought of as an ordered pair of real numbers (x, y). We shall be using the former form as it is more convenient in actual calculations. 'x' is called as the real part of z and will be written as $x = \text{Re } z$; while y is the imaginary part of z and will be written as $y = \text{Im } z$. (Note that by definition the imaginary part of z is y, and not iy, and is a real number).

$$z = \text{Re } z + i \text{ Im } z.$$

The complex conjugate of z is $z^* = x - iy = \text{Re } z - i \text{ Im } z$.

A convenient way to exhibit a complex number geometrically is via what is called an "Argand Diagram". This diagram uses the correspondence between the coordinates of a point in a plane and the ordered pair (x, y) corresponding to a complex number $z = x + iy$.

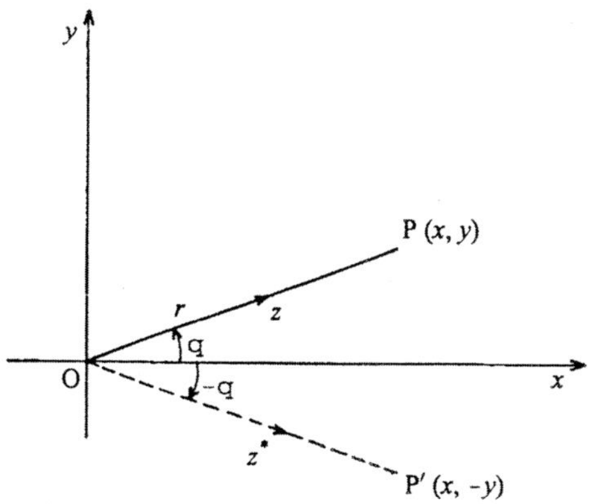

Fig. 6.1 *Argand diagram*

Consider a complex number $z = x + iy$. In the x–y plane plot the point P with coordinates (x, y). Then the complex number can be represented by the straight line drawn form O, the orgin, to P.

The length of OP, r, is called the magnitude or modulus of z and is denoted by $|z|$. Thus, $|z| = r = \sqrt{x^2 + y^2} = \sqrt{(\text{Re } z)^2 + (\text{Im } z)^2}$

Now, let OP make an angle θ with positive x-axis. Then

$$z = x + iy = r \cos\theta + ir \sin\theta \equiv re^{i\theta}. \tag{6.1}$$

(We have not yet defined exponential of a complex number. This will be done in §6.4.1 and the properties of $e^{i\theta}$ will be derived there. However, we may treat Eq. (6.1) as the definition of $e^{i\theta}$ and then the properties used follow from trigonometric identities). θ is called the argument of z and is determined only up to an addition of an angle $2\pi n$ where n is an integer (positive, negative or zero).

Consider the point P′ with coordinates $(x, -y)$. Then OP′ represents

$$z^* = x - iy = re^{-i\theta}.$$

The polar representation $z = re^{i\theta}$ is very useful under many circumstances, in particular in obtaining the following results:

(i)
$$z_1 z_2 = r_1 e^{i\theta_1} r_2 e^{i\theta_2} = r_1 r_2 e^{i(\theta_1+\theta_2)}$$
$$|z_1 z_2| = r_1 r_2 = |z_1|\,|z_2|.$$
$$\arg z_1 z_2 = \theta_1 + \theta_2 = \arg z_1 + \arg z_2$$

(The first line follows from the definition of $e^{i\theta}$ and trigonometric identities).

(ii)
$$zz^* = re^{i\theta} re^{-i\theta} = r^2 = |z^2|$$

i.e., $$|z^2| = zz^* = |z^*|^2.$$

(iii) $$\frac{1}{z} = \frac{1}{re^{i\theta}} = \frac{1}{r}e^{-i\theta} = \frac{re^{-i\theta}}{r^2} = \frac{z^*}{|z^2|}$$

(iv) $$\frac{z_1}{z_2} = r_1 e^{i\theta} \frac{1}{r_2} e^{-i\theta_2} = \frac{r_1}{r_2} e^{i(\theta_1-\theta_2)}$$

6.1.1 Triangle inequalities

Further results are obtained by using the triangle equalities on an Argand diagram in which one associates complex numbers with the three sides of a triangle (see Fig. 6.2).

Consider two complex numbers $z_1 = x_1 + iy_1$ and $z_2 = x_2 + iy_2$. Plot the points (x_1, y_1) and (y_2, y_2) at P_1 and P_2 respectively on the Argand plot. OP_1 and OP_2 represent z_1 and z_2 respectively. $z_3 \equiv z_1 + z_2 = (x_1 + x_2) + i(y_1 + y_2)$ is represented by OP_3 where P_3 is the point $(x_1 + x_2, y_1 + y_2)$. $OP_1P_3P_2O$ is a parallelogram. [$\mathbf{OP}_3 = \mathbf{OP}_1 + \mathbf{OP}_2$ in the usual vector language also].

Consider, now, the triangle OP_1P_3. We have the triangle inequalities

$$\begin{aligned} OP_3 &\le OP_1 + P_1P_3 \\ &= OP_1 + OP_2 \end{aligned} \tag{6.2}$$

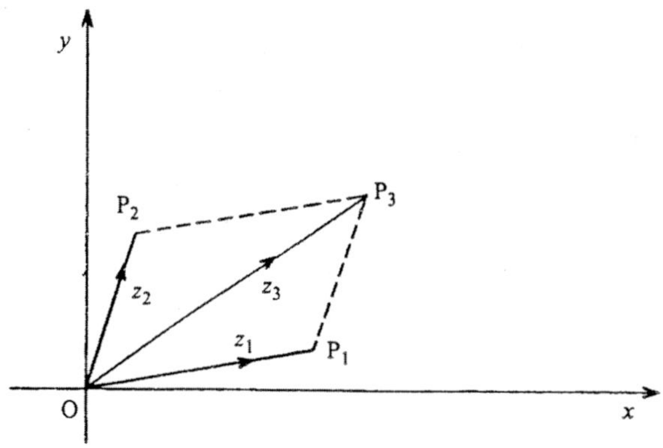

Fig. 6.2 *Triangle inequalities via Argand diagram*

$$\text{and} \quad |OP_1 - P_1P_3| = |OP_1 - OP_2| \leq OP_3 \tag{6.3}$$

But $OP_i = |z_i|$; $i = 1, 2, 3$. Hence, (6.2) and (6.3) lead to

$$|z_3| = |z_1 + z_2| \leq |z_1| + |z_2|$$

$$\text{and} \quad ||z_1| - |z_2|| \leq |z_3| = |z_1 + z_2|.$$

$$\text{Thus,} \quad ||z_1| - |z_2|| \leq |z_1 + z_2| \leq |z_1| + |z_2|.$$

Solved problem 6.1 Verify directly the triangle inequalities

$$||z_1| - |z_2|| \leq |z_1 + z_2| \leq |z_1| + |z_2|.$$

Solution: We shall work out the first part of the inequality and leave the second part to the reader, as it may be established in a quite analogous manner.

To show,

$$||z_1| - |z_2|| \leq |z_1 + z_2|$$

is equivalent to showing

$$[|z_1| - |z_2|]^2 \leq |z_1 + z_2|^2 = [z_1 + z_2]^*[z_1 + z_2]$$

$$\Rightarrow \quad |z_1|^2 + |z_2|^2 - 2|z_1|\,|z_2| \leq |z_1|^2 + |z_2|^2 + z_1^* z_2 + z_1 z_2^*$$

$$\Rightarrow \quad -2|z_1|\,|z_2| \leq (z_1^* z_2) + (z_1^* z_2)^*$$

$$\Rightarrow \quad -2|z_1^* z_2| \leq 2\,\text{Re}\,(z_1^* z_2)$$

$$\Rightarrow \quad |z_1^* z_2| \geq -\,\text{Re}\,(z_1^* z_2)$$

$$\Rightarrow \quad |-z_1^* z_2| \geq \text{Re}\,(-z_1^* z_2).$$

But this is always true as $|z| \geq \text{Re}\,(z)$ for any complex number.

6.2 Functions of a Complex Variable and Differentiability

A (scalar) function $f(x)$ of a real variable gives us a number f for each x at which f is defined. We want to generalize this definition to the function $f(z)$ of a complex variable z. We would naturally expect $f(z)$ to give a number, generally complex, for each z at which f is defined. Now we can also choose to look upon this $f(z)$ in terms of the usual real functions of two real variables x and y. For each z there is an ordered pair (x, y) and f associates with this pair of variables a number (generally complex), i.e., a collection of two real numbers Re f and Im f. Thus, a function f of a complex variable is obtained by putting together two real function $u(x, y)$ and $v(x, y)$ of two real variables x, y as

$$f(z) = u(x, y) + iv(x, y), \qquad u = \text{Re } f, v = \text{Im } f.$$

We next want to generalize the definitions of continuity and differentiability to functions of a complex variable.

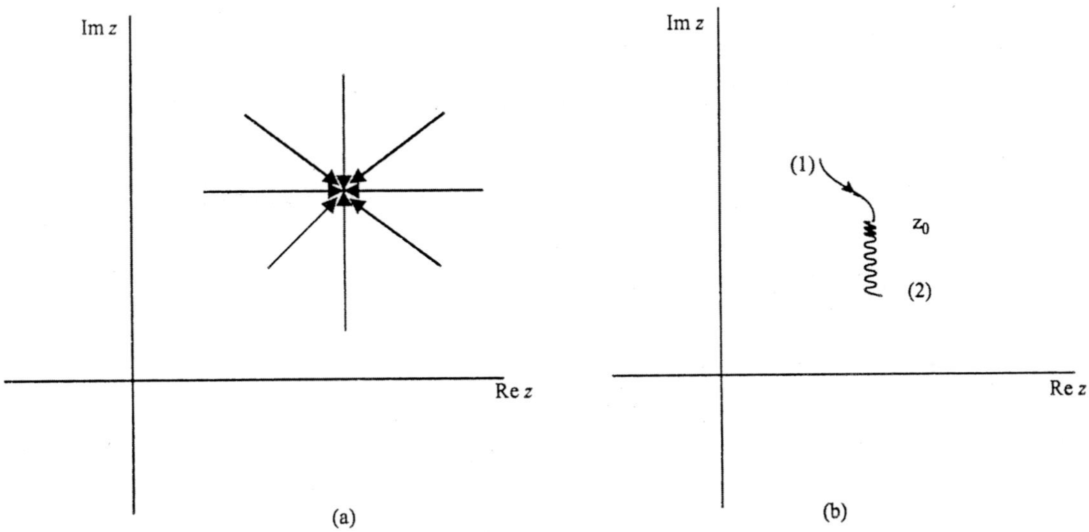

Fig. 6.3 *Definition of the derivative of a function of complex variable*

We will consider the limit of the quantity $\dfrac{f(z) - f(z_0)}{z - z_0}$ in order to possibly generalize the definition of the derivative of $f(z)$ at $z = z_0$. In the case of functions of a real variable, we had to consider a similar quantity $\dfrac{f(x) - f(x_0)}{x - x_0}$; its limit as $x \to x_0$ could be taken only in two directions $x \to x_0^+, x \to x_0^-$, i.e., ($x \to x_0$ from the right and from the left). These two limits had to exist and had to be identical in order that we could say that $f'(x_0)$ exists. In the case of functions of a complex variables, the limit $z \to z_0$ could be taken in infinitely many ways. For example, the point z_0 could be approached along any ray as shown in Fig. 6.3(a). It could also be approached along any curve say the curve (1) or (2) in Fig. 6.3(b). In order for $f'(z_0)$ to exist the limit $\dfrac{f(z)-f(z_0)}{z-z_0}$ taken in any of these ways must exist and yield the same answer. This requirement is highly nontrivial in general, as can be seen from some examples below.

Example Let $f(z) = x + 2iy$; and let $z_0 = 0$. So we want to test the existence of $f'(0)$. So we consider limit of $\dfrac{f(\Delta z) - f(0)}{\Delta z - 0}$ as $z \to 0$ along different direction.

(i) $z \to 0$ along the x-axis. Here $y = 0$; $\Delta z = \Delta x = x \equiv \varepsilon$.

$$\frac{f(\Delta z) - f(0)}{\Delta z - 0} = \frac{\varepsilon - 0}{\varepsilon - 0} = 1 \quad \text{for any } \varepsilon.$$

In particular,

$$\lim_{\Delta z \to 0} \frac{f(\Delta z) - f(0)}{\Delta z - 0} = 1 \quad \text{along the } x\text{-axis.}$$

(ii) $z \to 0$ along y-axis. Here $x = 0$; $\Delta z = i\Delta y \equiv iy \equiv i\delta$; $f(\Delta z) = 2i\delta$

$$\frac{f(\delta z) - f(0)}{\Delta z - 0} = \frac{2i\delta}{i\delta} = 2 \quad \text{for any } \delta.$$

In particular,

$$\lim_{\Delta z \to 0} \frac{f(\Delta z) - f(0)}{\Delta z - 0} = 2 \quad \text{along } y\text{-axis.}$$

Clearly these two limits differ from each other. In fact one can calculate the limit along the straight line $y = \alpha x$ with α arbitrary (but real). Here, $\Delta z = \Delta x + i\Delta y = \Delta x + i\alpha\Delta x \equiv (1 + i\alpha)\varepsilon$. $f(\Delta z) = \Delta x + 2i\Delta y = (1 + 2i\alpha)\varepsilon$.

Then, $$\lim_{\varepsilon \to 0} \frac{f(\Delta z) - f(0)}{\Delta z - 0} = \lim_{\varepsilon \to 0} \frac{(1 + 2i\alpha)\varepsilon}{(1 + i\alpha)\varepsilon} = \frac{1 + 2i\alpha}{1 + i\alpha}$$

and takes infinitely many different values depending on α.

Note: In this particular simple example, taking the limit $\varepsilon \to 0$ turned out to be redundant. But this is generally not the case.

6.2.1 Cauchy–Riemann conditions

We shall now derive a set of very important relations known as Cauchy–Riemann conditions (henceforth CR conditions) which are *necessary* for a function to be differentiable. We shall, later on, discuss the sufficient conditions for differentiability in Section 6.2.3.

Consider a function $f(z)$ defined over a domain D.

$$f(z) = u(x, y) + iv(x, y) \qquad z \in D.$$

Let us assume that $f(z)$ is differentiable at $z = z \in D$. Then $f'(z_0)$ exists. This means that

$$\lim_{z \to z_0} \frac{f(z) - f(z_0)}{z - z_0}$$

exists and is unique, i.e., independent of manner in which z approaches z_0. In particular, it must have the same value as $z \to z_0$ in the following two different directions (i) parallel to x axis and (ii) parallel to y axis (see Fig. 6.4).

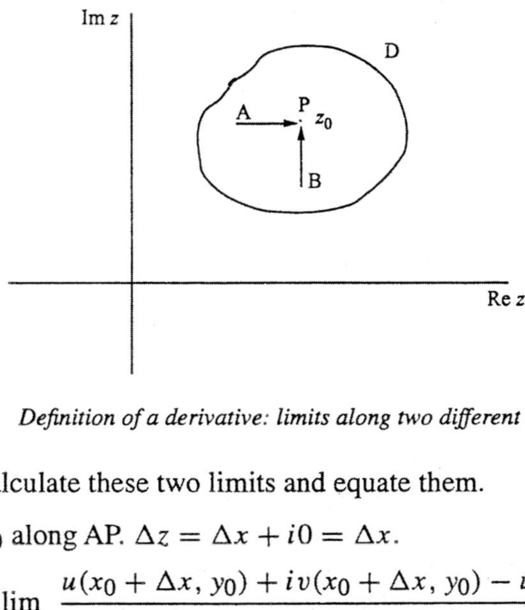

Fig. 6.4 *Definition of a derivative: limits along two different directions*

We shall therefore calculate these two limits and equate them.

(i) z approaches z_0 along AP. $\Delta z = \Delta x + i0 = \Delta x$.

$$\begin{aligned} f(z_0) &= \lim_{x \to x_0} \frac{u(x_0 + \Delta x, y_0) + iv(x_0 + \Delta x, y_0) - u(x_0, y_0) - iv(x_0, y_0)}{\Delta x} \\ &= \lim_{x \to x_0} \frac{u(x_0 + \Delta x, y_0) - u(x_0, y_0)}{\Delta x} \\ &\quad + i \lim_{x \to x_0} \frac{v(x_0 + \Delta x, y_0) - v(x_0, y_0)}{\Delta x}. \\ &= \left[\frac{\partial u}{\partial x} + i\frac{\partial v}{\partial x}\right] \text{evaluated at } x = x_0, y = y_0. \end{aligned} \tag{6.4}$$

In a similar manner we can calculate this limit along BP. Here $\Delta z = i\Delta y$. We then obtain

$$f'(z_0) = \left[-i\frac{\partial u}{\partial y} + \frac{\partial v}{\partial y}\right] \text{ evaluated at } x = x_0,\ y = y_0. \tag{6.5}$$

The two expressions (6.4) and (6.5) must, in particular, agree as $f'(z_0)$ exists.

Equating the real and imaginary parts of (6.4) and (6.5) leads us to

$$\frac{\partial u}{\partial x} = \frac{\partial v}{\partial y}; \qquad \frac{\partial u}{\partial y} = -\frac{\partial v}{\partial x}.$$

Or, in short, in obvious notations,

$$u_x = v_y \text{ and } u_y = -u_x.$$

The above two equations are referred to as the Cauchy–Riemann conditions.

Example 1 Consider the function $f(z) = z^2 \quad z \in \mathbb{C}$

Here, $u(x, y) = x^2 - y^2; \quad v(x, y) = 2xy$.

For $f(z)$ to be differentiable, CR conditions must necessarily be obeyed. So we calculate

$$u_x = 2x;\ u_y = -2y;\ v_x = 2y;\ v_y = 2x.$$

Thus the CR conditions $u_x = v_y$ and $u_y = -v_x$ are obeyed everywhere and these *necessary* conditions for differentiability of $f(z)$ are fulfilled everywhere. (This, by itself, does not however mean that $f(z)$ is differentiable; though, as we shall see later, it is so.)

Example 2 Let $f(z) = |z|^2 = x^2 + y^2$

$$u = x^2 + y^2;\ v = 0$$
$$u_x = 2x;\ u_y = 2y;\ v_x = 0 = v_y.$$

$u_x = v_y;\ u_y = -v_x$ is not satisfied unless $x = y = 0$, i.e., $z = 0$. Hence the function $|z|^2$ is not differentiable at any point except possibly the origin $z = 0$.

(Again the fact that CR conditions are satisfied at $z = 0$ by itself cannot say whether $f(z)$ is differentiable at $z = 0$. However, we shall examine this in the light of sufficient conditions for differentiability and show that this happens to be true.)

Example 3 Let $f(z) = x + 2iy$. (This is the example considered earlier.)

$$u = x;\ v = 2y;$$
$$u_x = 1;\ u_y = 0;\ v_x = 0;\ v_y = 2;$$
$$u_x \neq v_y \text{ for any } z.$$

Hence $f(z)$ is nowhere differentiable. (We have, of course, seen that $f'(z)$ did not exist at $z = 0$ in an earlier example).

6.2.2 CR conditions in terms of independent variables z and z^*

CR conditions take a very simple and easy-to-handle form when the independent variables are chosen as z and z^*. To this end, we note:

$$x = (z + z^*)/2; \qquad y = (z - z^*)/2i. \tag{6.6}$$

Now consider any function $f(z) = u(x, y) + iv(x, y)$. We rewrite this by substituting for x and y in the arguments of u and v from (6.6) as:

$$f(z) = u\left[\frac{z+z^*}{2}, \frac{z-z^*}{2i}\right] + iv\left[\frac{z+z^*}{2}, \frac{z-z^*}{2i}\right] \equiv g(z, z^*).$$

The R.H.S is now expressed as a function g of two independent variables z and z^* rather than of x and y, viz. $g(z, z^*)$.†

Let us now calculate $\frac{\partial}{\partial z^*} g(z, z^*)$. (Here, we mean to vary z^* keeping z fixed). By chain rule of differentiation,

$$\begin{aligned} \frac{\partial}{\partial z^*} g(z, z^*) &= \frac{\partial x}{\partial z^*}\frac{\partial g}{\partial x} + \frac{\partial y}{\partial z^*}\frac{\partial g}{\partial y} \\ &= \frac{1}{2}\frac{\partial}{\partial x}(u + iv) - \frac{1}{2i}\frac{\partial}{\partial y}(u + iv) \\ &= \frac{1}{2}[u_x - v_y] + i\frac{1}{2}[v_x + u_y]. \end{aligned} \tag{6.7}$$

Thus from (6.7) it is seen that the complex quantity $\frac{\partial}{\partial z^*} g(z, z^*)$ vanishes if CR conditions are fulfilled. Thus, the CR conditions are easily expressed by a single complex equation

$$\frac{\partial}{\partial z^*} g(z, z^*) = 0.$$

This formulation of CR conditions is at times more convenient. To see this, consider again the three examples just dealt with in the last subsection:

(i) $f(z) = z^2 = g(z, z^*)$, $\dfrac{\partial}{\partial z^*} g(z, z^*) = 0$.
CR conditions are satisfied everywhere.

(ii) $f(z) = |z|^2 = zz^* = g(z, z^*)$. Now $\dfrac{\partial}{\partial z^*} g = z = 0$ only at origin.

†Here we clarify a point. One may find it absurd to write an equation $f(z) = g(z, z^*)$ where f seems to be a function of only one variable z. This apparent contradiction has arisen on account of different conventions used in writing arguments for functions f and g. Recall the original definition of a function of a complex variable (Section. 6.2) and note that $f(z)$ is meant to be a function of two independent real variables x and y; these are entirely equivalent to the variables z and z^*.

(iii) $$f(z) = x + 2iy = \frac{z+z^*}{2} + 2i\frac{z-z^*}{2i} = \frac{3}{2}z - \frac{1}{2}z^* = g(z, z^*).$$

$$\frac{\partial g}{\partial z^*} = -\frac{1}{2} \text{ is nowhere zero.}$$

These examples justify the utility of this formulation.

6.2.3 Sufficient conditions for differentiability

An emphasized earlier, CR conditions are necessary conditions for differentiability at a point. These are enough to conclude that a function is not differentiable at $z = z_0$ if they (CR conditions) are not satisfied at $z = z_0$. But when one wants to find the set of points for which $f(z)$ is differentiable, CR conditions are not sufficient to guarantee this. This is borne out by the example given below.

Example We shall now give an example of a function for which the partial derivatives u_x, u_y, v_x, v_y all exist at a point (x_0, y_0), satisfy CR conditions and are continuous there, but theses derivatives do not exist in any neighbourhood of (x_0, y_0).

Consider $f(z) = u(x, y) + iv(x, y)$ with

$$\begin{aligned} u(x, y) &= x + y \qquad && x = 0 \text{ and/or } y = 0 \\ &= 1 && \text{otherwise} \end{aligned}$$

and

$$\begin{aligned} v(x, y) &= -x + y \qquad && x = 0 \text{ and/or } y = 0 \\ &= 1 && \text{otherwise.} \end{aligned}$$

It is easily seen that

$$\begin{aligned} u_x(0, 0) &= 1 = u_y(0, 0) \\ -v_x(0, 0) &= 1 = v_y(0, 0) \end{aligned}$$

so that the CR conditions are satisfied at (0, 0).

But $u_x(0, \varepsilon_2)$, $u_y(\varepsilon_1, 0)$, $v_x(0, \varepsilon_2)$ and $v_y(\varepsilon_1, 0)$ do not exist for *any* $\varepsilon_1 \neq 0$ and any $\varepsilon_2 \neq 0$. Hence every disc $|x + iy| < \delta$ contains points where partial derivatives do not exist.

It is also seen that the function is not differentiable at (0, 0). For, say along the line $x = y$,

$$\lim_{\substack{x \to 0 \\ y \to 0}} \frac{f(x, y) - f(0, 0)}{x + iy} = \frac{1 - 0}{x + iy}$$

does not exist. We shall see that this example does not fulfil the sufficient conditions for differentiability.

We now formulate a set of sufficient conditions for differentiability.

Theorem 6.1 A function $f(z) = u(x, y) + iv(x, y)$ is differentiable at $z = z_0 = x_0 + iy_0$ within the domain D of its definition if there exists a neighbourhood of (x_0, y_0) throughout which u_x, u_y, v_x, v_y exist and further these are continuous at (x_0, y_0) and satisfy CR conditions there.

Proof: Let u_x, u_y, v_x, v_y exist in the circle $|z - z_0| < \in$, and be continuous at (x_0, y_0). For Δx and $\Delta y < \varepsilon$, consider

$$\begin{aligned} &u(x_0 + \Delta x, y_0 + \Delta y) - u(x_0, y_0) \\ &= u(x_0 + \Delta x, y_0 + \Delta y) - u(x_0, y_0 + \Delta y) + u(x_0, y_0 + \Delta y) - u(x_0, y_0). \end{aligned} \tag{6.8}$$

Now as $u_x(x_0, y_0 + \Delta y)$ exists,

$$\lim_{\Delta x \to 0} \frac{u(x_0 + \Delta x, y_0 + \Delta y) - u(x_0, y_0 + \Delta y)}{\Delta x} = u_x(x_0, y_0 + \Delta y), \tag{6.9}$$

and this implies that

$$\frac{u(x_0 + \Delta x, y_0 + \Delta y) - u(x_0, y_0 + \Delta y)}{\Delta x} - u_x(x_0, y + \Delta y) \equiv \delta_1' \tag{6.10}$$

is such that $|\delta_1'|$ can be made arbitrarily small by making Δx sufficiently small.

Thus,

$$u(x_0 + \Delta x, y_0 + \Delta y) - u(x_0, y_0 + \Delta y) = u_x(x_0, y_0 + \Delta y)\Delta x + \delta_1' \Delta x. \tag{6.11}$$

In a similar manner

$$u(x_0, y_0 + \Delta y) - u(x_0, y_0) = \left[u_y(x_0, y_0) + \delta_2'\right] \Delta y \tag{6.12}$$

with $\delta_2' \to 0$ as $\Delta y \to 0$.

Further, as $u_x(x, y)$ is continuous at (x_0, y_0),

$$u_x(x_0, y_0 + \Delta y) - u_x(x_0, y_0) = \delta_1''(\Delta y) \tag{6.13}$$

with $\delta_1'' \to 0$ and $\Delta y \to 0$.

Here (6.13) becomes

$$\begin{aligned} u(x_0 + \Delta x, y + \Delta y) - u(x_0, y_0 + \Delta y) &= \left[u_x(x_0, y_0) + \delta''_1 + \delta_1'\right] \Delta x \\ &= \left[u_x(x_0, y_0) + \delta_1\right] \Delta x \end{aligned} \tag{6.14}$$

with $\delta_1 \equiv \delta_1' + \delta_1''$ and $\delta_1 \to 0$ and $\Delta x \to 0$, $\Delta y \to 0$.

Adding (6.14) and (6.11) we have

$$u(x_0 + \Delta x, y + \Delta y) - u(x_0, y_0) = u_x(x_0, y_0)\Delta x + u_y(x_0, y_0)\Delta y \\ + \delta_1 \Delta x + \delta_2 \Delta y \tag{6.15}$$

with $\delta_1 \to 0$ and $\delta_2 \to 0$ as $\Delta x \to 0$ and $\Delta y \to 0$.

In a similar manner,

$$\begin{aligned} i[v(x_0 + \Delta x, y_0 + \Delta y) - v(x_0, y_0)] &= i[v_x(x_0, y_0)\Delta x + v_y(x_0, y_0)\Delta y \\ &\quad + \varepsilon_1 \Delta x + \varepsilon_2 \Delta y] \\ &= i[-u_y(x_0, y_0)\Delta x + u_x(x_0, y_0)\Delta y + \varepsilon_1 \Delta x + \varepsilon_2 \Delta y] \end{aligned} \tag{6.16}$$

in view of the CR conditions at (x_0, y_0). Here $\varepsilon_1, \varepsilon_2 \to 0$ as Δx and $\Delta y \to 0$.

Adding (6.15) and (6.16),

$$f(z_0 + \Delta z) - f(z_0) = u_x(x_0, y_0)\Delta z - i u_y \Delta z \\ + (\delta_1 + i\varepsilon_1)\Delta x + (\delta_2 + i\varepsilon_2)\Delta y.$$

Thus

$$\begin{aligned} \left| \frac{f(z_0 + \Delta z) - f(z_0)}{\Delta z} - u_x + i u_y \right| &= \left| (\delta_1 + i\varepsilon_1)\frac{\Delta x}{\Delta z} + (\delta_2 + i\varepsilon_2)\frac{\Delta y}{\Delta z} \right| \\ &\leq |(\delta_1 + i\varepsilon_1)| \left| \frac{\Delta x}{\Delta y} \right| + |(\delta_2 + i\varepsilon_2)| \left| \frac{\Delta y}{\Delta z} \right| \\ &\leq \sqrt{\delta_1^2 + \varepsilon_1^2} + \sqrt{\delta_2^2 + \varepsilon_2^2} \\ &\to 0 \text{ as } \Delta z = \Delta x + i\Delta y \to 0. \end{aligned}$$

(Note that this is irrespective of how $\Delta z \to 0$).

Hence the following limit

$$\lim_{\Delta z \to 0} \frac{f(z_0 + \Delta z) - f(z_0)}{\Delta z} = u_x - i u_y$$

exists. So $f'(z_0)$ exists and is given by

$$f'(z_0) = u_x - i u_y = u_x + i v_x.$$

6.2.4 CR conditions in polar coordinates

Often, it is convenient to express functions of a complex variable in its polar form, i.e., in terms of r and θ (instead of x and y). For example,

$$f(z) = z^{1/3} \equiv r^{1/3} e^{i\theta/3} \tag{6.17}$$

is much more neatly expressed in terms of r and θ rather than as

$$f(z) = \sqrt[6]{x^2 + y^2} \left[\cos\left(\frac{1}{3} \tan^{-1} \frac{y}{x} \right) + i \sin\left(\frac{1}{3} \tan^{-1} \frac{y}{x} \right) \right]. \tag{6.18}$$

Testing the function for differentiability using the latter form (6.18) via CR conditions would be very tedious. While dealing with such functions, it would be much more desirable to have CR conditions cast in terms of polar coordinates. This is easily done as follows. Let

$$f(z) = u(x, y) + iv(x, y) \equiv u(r, \theta) + iv(r, \theta).$$

(Here we clarify that $u(r, \theta)$ is *numerically* same as $u(x, y)$ at a given point. The *functional forms* of $u(r, \theta)$ and of $u(x, y)$ are not the same despite our convention for this notation). We now express partial derivatives of u and v with respect to r and θ in terms of u_x, u_y, v_x, v_y. To do this, use the chain rule of differentiation (and $x = r\cos\theta$, $y = r\sin\theta$).

$$\begin{aligned}
\frac{\partial u}{\partial r} &= \frac{\partial x}{\partial r} u_x + \frac{\partial y}{\partial r} u_y = \frac{x}{r} u_x + \frac{y}{r} u_y & r \neq 0 \\
\frac{\partial u}{\partial \theta} &= \frac{\partial x}{\partial \theta} u_x + \frac{\partial y}{\partial \theta} u_y = -y u_x + x u_y & \\
\frac{\partial v}{\partial r} &= \frac{x}{r} v_x + \frac{y}{r} v_y & r \neq 0 \\
\frac{\partial v}{\partial \theta} &= -y v_x + x v_y &
\end{aligned} \tag{6.19}$$

Now, if CR conditions are satisfied, we can express $\frac{\partial v}{\partial r}$ in terms of u_y and u_x to find

$$\frac{\partial v}{\partial r} = \frac{-x u_y + y u_x}{r} = -\frac{1}{r} \frac{\partial u}{\partial \theta} \qquad (r \neq 0) \tag{6.20}$$

and similarly we find

$$\frac{\partial v}{\partial \theta} = y u_y + x u_x = r \frac{\partial u}{\partial r}. \tag{6.21}$$

The converse is also easy to verify. If Eqs. (6.20–21) are satisfied, CR conditions in Cartesian form are also satisfied. Further, except for the point $r = 0$, existence u_x and u_y implies the existence of $\frac{\partial u}{\partial r}$ and $\frac{\partial u}{\partial \theta}$. The converse is also easily established. Existence of $\frac{\partial u}{\partial r}, \frac{\partial u}{\partial \theta}, \frac{\partial v}{\partial r}, \frac{\partial v}{\partial \theta} (r \neq 0)$ implies the existence of u_x, u_y, v_x, v_y for $z \neq 0$. Thus at $r \neq 0$, the CR conditions in Cartesian form

(i) u_x, u_y, v_x, v_y exist and satisfy $u_x = v_y$ and $u_y = v_x$
are equivalent to the CR conditions in polar form given below.

(ii) $u_r, u_\theta, v_r, v_\theta$ exist and satisfy

$$u_r = \frac{1}{r}\frac{\partial v}{\partial \theta} \quad \text{and} \quad v_r = -\frac{1}{r}\frac{\partial u}{\partial \theta}.$$

Further, the derivative $f'(z)$ can be written as $f'(z) = (u_r + iv_r)e^{-i\theta}$ (see Problem 6.15).

Example Test the function $f(z) = z\,\bar{z} = r^2$ for differentiability.

Here $u = r^2$; $v = 0$; $u_r = 2r$, $u_\theta = 0 = v_r = v_\theta$. $u_r = \frac{1}{r}v_\theta$ is satisfied nowhere for $r \neq 0$. Hence the function cannot be differentiable anywhere except possibly at the origin. (The derivation of CR conditions in polar coordiante given above by itself does not allow for testing for differentiability at $r = 0$).

Sufficient conditions for diffferentiability in polar form

The sufficient conditions for differentiability of a function stated in Section 6.2.3 in terms of the variables (x, y) can also be cast in terms of the variables r and θ. To this end, we use the relations (6.19). Then it is easy to see that the sufficient conditions of Section 6.2.3 translate into the result:

A function $f(z) = u(r, \theta) + iv(r, \theta)$ is differentiable at $z_0 = re^{i\theta}$ $(r \neq 0)$ within the domain D of its definition if there exists a neighbourhood of z_0 throughout which $u_r, u_\theta, v_r, v_\theta$ exist, and these are further continuous at z_0 and satisfy the CR conditions at z_0.

Example Find the region of differentiability of the function

$$f(z) = r^{1/3}e^{i\theta/3} \qquad 0 \leq \theta < 2\pi.$$

Here

$$u = r^{1/3}\cos\frac{\theta}{3}; \qquad v = r^{1/3}\sin\frac{\theta}{3};$$

$$u_r = \frac{1}{3}r^{-2/3}\cos\frac{\theta}{3}; \quad v_r = \frac{1}{3}r^{-2/3}\sin\frac{\theta}{3} \qquad (r \neq 0);$$

$$u_\theta = -\frac{1}{3}r^{1/3}\sin\frac{\theta}{3}; \quad v_\theta = \frac{1}{3}r^{1/3}\cos\frac{\theta}{3} \qquad (0 < \theta < 2\pi).$$

Note that $u_r = \frac{1}{r}v_\theta$ and $v_r = -\frac{1}{r}u_\theta$ are satisfied for $r \neq 0$ and $0 < \theta < 2\pi$. Also all first order partial derivatives are continous in the domain. So the function is differentiable at every point in the domain $r \neq 0$, $0 < \theta < 2\pi$. But it is easily seen that neither u nor v are continuous at across the positive real axis. (Hence, u_θ and v_θ have been given above only for $0 < \theta < 2\pi$). Hence, as the function is not even continuous on positive real axis, it is not differentiable there.

Further, the derivative $f'(z)$ in the domain of differentiability is given by

$$\begin{aligned} f'(z) &= e^{-i\theta}(u_r + iv_r) \\ &= e^{-i\theta}\frac{1}{3}r^{-2/3}(\cos\theta/3 + i\sin\theta/3) \\ &= \frac{1}{3}r^{-2/3}e^{-2i\theta/3} \qquad 0 < \theta < 2\pi;\ r \neq 0. \end{aligned}$$

6.3 Analytic Functions

We have already defined what we mean by differentiability of a function $f(z)$ at a given point z_0 in its domain of definition. As we have seen, there are examples of functions of a complex vairable which are differentiable only at some isolated points (e.g., $f(z) = |z|^2$ is differentiable only at $z = 0$). This, then, shows that the concept of differentiability is "too broad" a concept and we will need to introduce a somewhat narrower concept of "analyticity" here. The class of analytic functions will turn out to be a very important class of functions of a complex variable and many interesting and beautiful results can be derived for these. In particular, most of our furture discussion will be based on functions that are analytic in some domain.

Definition: A function of a complex variable $f(z)$ is analytic at z_0 in the domain of its definition D if there exists a neighbourhood $|z - z_0| < \varepsilon$, $(\varepsilon > 0)$ in D, at *every* point of which $f(z)$ is differentiable.

Thus, analyticity of $f(z)$ at a point z_0 requires differentiability in a whole neighbourhood of z_0, however small.

Note: In this case, the definition implies that $f(z)$ is analytic not only at z_0, but happens to be analytic in the entire neighbourhood $|z - z_0| < \varepsilon$. (See Problem 6.13.) Thus, analyticity of a function is a property intrinsically referring to a region (however small) rather that a point.

Examples

(1) Let $f(z) = \frac{1}{z-1} (z \neq 1)$. It can be easily verified that $f(z)$ is differentiable everywhere except at $z = 1$. $\left(\text{Note in particular: } \frac{\partial}{\partial z^*}\left(\frac{1}{z-1}\right) = 0,\ z \neq 1 \text{ including points arbitrarily close to } z = 1\right)$. Hence the function is analytic everywhere except at $z = 1$. Note that analyticity is valid at a point z_1 however close to (but not coincident with) $z = 1$. This is so because one can find a neighbourhood of z_1, not containing $z = 1$, throughout which the function is differentiable.

(2) $f(z) = |z|^2$: As remarked earlier the function is differentiable nowhere except at $z = 0$. So it cannot be analytic for $z \neq 0$. It is not analytic at $z = 0$ either, because every neighbourhood at $z = 0$ contains points where $f(z)$ is not differentiable.

(3) $f(z) = z + az^2 + bz^3$.
$f(z)$ is differentiable everywhere, with $f'(z) = 1+2az^2+3bz^3$. Hence f is analytic everywhere in the entire complex plane.

Definition:

Entire function: A function that is analytic in the entire complex plane is called an entire function. The function $f(z)$ in example (3) above is entire. In fact every polynomial in z is an entire function.

Singularity: If a function $f(z)$ is not analytic at $z = z_0$ but is analytic at some point in every neighbourhood of $z = z_0$. then $z = z_0$ is a singular point of $f(z)$.

In the example given under sufficient condition for differentiability in polar form, $f(z) = \frac{1}{z-1}$ has a singular point at $z = 1$. This is so because it is not analytic at $z = 1$, yet in any neighbourhood of $z = 1$ there are points of analyticity of $f(z)$ (in this example, all the remaining points of the neighbourhood, in fact, are such points).

Theorem 6.2 If two functions f and g are analytic in D, their sum $(f + g)$ and product f, g are also analytic in D.

Proof: The proofs are straightforward. If f and g are differentiable at $z \varepsilon D$, so is $f + g$ and fg with

$$\frac{d}{dz}(f+g) = \frac{df}{dz} + \frac{dg}{dz}$$

and
$$\frac{d}{dz}(f \cdot g) = f\frac{df}{dz} + \frac{dg}{dz}g.$$

Thus, both $f + g$ and fg are differentiable for every $z \in D$ and hence are both analytic in D.

Theorem 6.3 If two functions f and g are both analytic in D and $g(z) \neq 0$ for any $z \in D$, then $f(z)/g(z)$ is analytic in D.

This follows because f/g is differentiable at every point in D with its derivative given by

$$\frac{d}{dz}\left(\frac{f}{g}\right) = \frac{1}{g}\frac{df}{dz} - \frac{1}{g^2}\frac{dg}{dz}f.$$

which is defined everywhere in D.

Theorem 6.4 Let f and g be entire. Then $h(z) = f[g(z)]$ is entire.

This follows from the existence of derivative of h at every $z \in \mathbb{C}$ and is given by

$$\frac{dh}{dz} = \left.\frac{df(\omega)}{d\omega}\right|_{\omega=g(z)} \frac{dg(z)}{dz},$$

each factor existing for $w \in \mathbb{C}$ and $z \in \mathbb{C}$, respectively.

A result such as the above one can be given also for functions that are not entire, but analytic in a domain, instead.

6.3.1 Harmonic functions

We shall now establish a simple property for the real and imaginary parts (u and v) of analytic function $f(z)$. This property is, in fact, the basis for extensive applications of the theory of analytic functions and conformal mapping to problems in electrostatics in two dimensions, as we will see later.

To establish this property, consider an analytic function $f(z) = u(x, y) + iv(x, y)$ over a domain D. Then in D, u and v satisfy CR condtions,

$$u_x = v_y, \qquad u_y = -v_x.$$

Now, assuming that the second order partial derivatives of u and v exist and are continuous at a point (as it turns out, this assumption is indeed unnecessary. See Section. 7.4), we obtain

$$u_{xx} = v_{yx}, \qquad u_{yy} = -v_{xy}.$$

Now, continuity of partial derivatives guarantees $v_{xy} = v_{yx}$. Thus, we have

$$u_{xx} + u_{yy} = 0.$$

Or, introducing the Laplacian in 2 dimensions, $\nabla^2 = \frac{\partial^2}{\partial x^2} + \frac{\partial^2}{\partial y^2}$, we have

$$\nabla^2 u(x, y) = 0 \quad \text{in } D.$$

A function satisfying this (Laplace's) equation in D is said to be harmonic in D. In a similar manner, the CR conditions lead to a similar result for $v(x, y)$:

$$\nabla^2 v(x, y) = 0 \quad \text{in } D.$$

Thus, both the real part $u(x, y)$ and the imaginary part $v(x, y)$ of an analytic function $f(z)$ are harmonic in the domain D of analyticity of $f(z)$.

The reason why analytic functions are useful in electrostatics problems in two dimensions is now evident. The electrostatic potential function $\phi(x, y)$ (in two-dimensional problems) in a region free of charge satisfies just the Laplace's equation:

$$\nabla^2\phi(x, y) = 0.$$

This suggests that the solution to the electrostatic potential problem may be (say) the real part $u(x, y)$ of an analytic function $f(z)$. In fact, the real part (and also the imaginary part) of an analytic function $f(z)$ indeed represents solution to *some* electrostatic problem. Now, this identification by itself would not be so helpful were it not for the following observation: suppose Re $f(z)$ represents solution to electrostatic potential of one problem. One can make an analytic transformation $z = g(\omega)$ and thus generate a function $f[g(\omega)]$ which is also analytic in ω. Then Re $f[g(\omega)]$ represents solution to another electrostatic problem. Which electrostatic problem Re $f[g(\omega)]$ refers to can be found out knowing the original problem and the analytic transformation $z = g(\omega)$. (For details, see Chapter 8.) Thus, by knowing the solution to the electrostatic potential in a simple case, it becomes possible to solve a much more complicated problem (and in fact as many problems as there are analytic transformation viz. infinite number of them, in principle). The above is only intended to give a qualitative idea of use of analytic functions in electrostatics. We leave the detailed discussion to Chapter 8.

Definition: If two functions u and v are harmonic in D and satisfy CR conditions throughout D, then v is said to be the harmonic conjugate to u in D.

The imaginary part of an analytic function is harmonically conjugate to its real part.

6.4 Special Analytic Functions

In this subsection we shall see how to generalize some special functions of a real variable x ($\sin x$, e^x, $\log x$, x^a, etc.) to analogous functions of a complex variable z. This is done by requiring that these generalization be analytic (in as large a domain as possible). We list a number of motivations for doing this.

(i) The first motivation is purely mathematical. It is natural to seek functions of a complex variable which reduce to the usual functions of a real variable on the real axis and are analytic in the complex plane. Students are already familiar with the use of such functions, and the convenience they lead to (e.g. $\cos\theta = \frac{e^{i\theta}-e^{-i\theta}}{2}$ involves exponential function $e^{i\theta}$ of a complex variable $i\theta$).

(ii) In next chapter we shall see that the evaluation of real integrals often becomes much simplified if they can be written as contour integrals over the complex plane. This is possible if the integral, which is a function of a real variable, can also be defined in the complex plane, i.e., as a function of z. This needs the said generalization of function to the complex plane.

(iii) In Chapter I of the accompanying volume of this book (*Mathematical Physics: Advanced Topics*), we shall see that the treatment of ordinary differential equations becomes greatly simplified if the independent variable is chosen to be complex instead of real. Thus, generalization of functions of a real variable naturally enter the solutions of such differential equations. For example, the equation

$$\frac{d^2y}{dx^2} + x = 0$$

has as a solution $y(x) = A\sin x$. The analogous complex equation

$$\frac{d^2y(z)}{dz^2} + z = 0$$

would have as a solution $y(z) = A\sin z$. This calls for a definition of $\sin z$ for z complex.

Now let, us consider this problem in more detail. Suppose we are given function $f(x) = e^x$. We are required to find a function of a complex variable $f(z)$ that generalizes this. Thus, $f(x + i0)$ must reduce to e^x. Firstly, we can ask, is $f(z)$ then unique? The answer, generally, is no. For, if $f_1(z)$ is one such generalization, $f_2(z) = f_1(z)+g(y)$ is also another one, where $g(y)$ is any function of y with the property $g(0) = 0$ (say $g(y) = \sin y$). Naturally, we need further restrictions. In view of the very powerful properties (some to be established in next chapter) enjoyed by analytic functions, it would be natural to restrict $f(z)$ to be analytic. And this restriction proves sufficient for our purpose. As shown in the Chapter 8 (Section. 8.1) on Analytic Continuation, there is a unique analytic function $f(z)$ that generalizes $f(x) = e^x$ to the entire complex plane.

6.4.1 The exponential functions

Here, we want to generalize the exponential function e^x to an analytic function $f(z)$ that yields $f(x + i0) = e^x$. As we have seen from our previous experience, constructing a function of z alone (i.e. having no z^* dependence) and obeying certain further conditions of existence and continuity of partial derivatives gives us an analytic function. Now, we therefore need the definition of $e^z \equiv e^{x+iy}$ so that e^z would reduce automatically to e^x for $y = 0$ and being a function of z alone, be analytic. (Note that we are not to assume here the definition of e^{iy}, i.e., exponential of a purely imaginary quantity as the argument iy is already complex and this would lead us to into circles). We proceed from the property for real variables x_1 and x_2, viz. $e^{x_1+x_2} = e^{x_1} \cdot e^{x_2}$ to make a factorization ansatz:

$$e^z = e^{x+iy} = e^x \, G(y) \tag{6.22}$$

with $G(0) = 1$. Expressing G in real and imaginary parts

$$e^z = e^x \, g_1(y) + i \, e^x g_2(y) = u + iv \tag{6.23}$$

with $g_1(0) = 1$, $g_2(0) = 0$. Now u satisfies the Laplace equation

$$u_{xx} + u_{yy} = 0 \tag{6.24}$$

which yields

$$\frac{d^2}{dy^2} g_1 + g_1 = 0 \quad \text{with boundary condition } g_1(0) = 1. \tag{6.25}$$

Further, the CR condition $u_y = -v_x$ yields

$$g_2(y) = -g_1'(y). \tag{6.26}$$

$$\text{Thus,} \quad g_1'(0) = 0. \tag{6.27}$$

It is now easy to see that the unique solution to (6.25), (6.26), (6.27) is

$$g_1(y) = \cos y, \quad g_2(y) = \sin y.$$

This thus yields us the definition

$$e^z = e^{x+iy} \equiv e^x(\cos y + i \sin y). \tag{6.28}$$

In particular, setting $x = 0$, we obtain the result (the Euler's formula)

$$e^{iy} = \cos y + i \sin y \tag{6.29}$$

which was introduced simply as a notation in Section. 6.1.

We further note $e^{x+iy} = e^x\, e^{iy} = e^{iy}e^x$.

Using the result $e^{iy_1}\, e^{iy_2} = e^{i(y_1+y_2)}$ established in Section 6.1, it'is easy to verify the additive property

$$e^{z_1}e^{z_2} = e^{x_1+iy_1}e^{x_2+iy_2} = e^{x_1}e^{iy_1}e^{x_2}e^{iy_2} = e^{x_1+x_2}e^{i(y_1+y_2)} = e^{z_1+z_2}.$$

In a similar manner,

$$e^{z_1}/e^{z_2} = e^{z_1-z_2}.$$

Putting $z_1 = 0$ and $z_1 = z$, this leads to

$$\frac{1}{e^z} = e^{-z}$$

Further, e^z is entire with derivative given by

$$\frac{d}{dr}(e^z) = u_x + iv_x = e^x \cos y + i\, e^x \sin y = e^z. \tag{6.30}$$

6.4.2 Trignometric functions

We shall now see how to generalize the trignometric functions of a real variable x to the respective analytic functions of the complex variable z. This turns out to be rather easy now that we have already generalized the exponential function. For, in the case of a real variable x,

$$\cos x = \frac{e^{ix} + e^{-ix}}{2}, \qquad \sin x = \frac{e^{ix} - e^{-ix}}{2i}. \tag{6.31}$$

Now, we know that e^{ix} can be generalized to an analytic function e^{iz} as defined in the previous section. We, thus, define

$$\cos z = \frac{e^{iz} + e^{-iz}}{2}, \qquad \sin z = \frac{e^{iz} - e^{-iz}}{2i}. \tag{6.32}$$

Since e^{iz} is entire (note the CR conditions $\frac{\partial}{\partial z^*}e^{iz} = 0$), so are $\cos z$ and $\sin z$. In view of the fact that (see Eq. (6.30) of Section 6.4.1) $\frac{d}{dz}e^{\pm iz} = \pm ie^{\pm iz}$, it is easily verified that

$$\frac{d}{dz}\sin z = \cos z; \quad \frac{d}{dz}\cos z = -\sin z. \tag{6.33}$$

From the defining relations, it is easy to verify directly (see Ex. 6.18)

$$\sin(z + 2n\pi) = \sin z \qquad \cos(z + 2n\pi) = \cos z \tag{6.34}$$

$$\sin(-z) = -\sin z \qquad \cos(-z) = \cos z \tag{6.35}$$

$$\sin(\pm z + \pi/2) = \cos z \qquad \cos(\pm z + \pi/2) = \mp \sin z \tag{6.36}$$

$$\sin(z + \pi) = -\sin z \qquad \cos(z + \pi) = -\cos z \tag{6.37}$$

and (see Ex. 6.19)

$$\sin(z_1 + z_2) = \sin z_1 \cos z_2 + \cos z_1 \sin z_2. \tag{6.38}$$

Setting $z_2 = \pi/2 - z$ and $z_1 = z$ in the above equation,

$$\sin \pi/2 = \sin z \cos(\pi/2 - z) + \cos z \sin(\pi/2 - z).$$

Using (6.36) in the above we get

$$1 = \sin^2 z + \cos^2 z. \tag{6.39}$$

Similarly, setting $z_2 = z_2' + \pi/2$ in (6.38),

$$\sin(z_1 + z_2' + \pi/2) = \sin z_1 \cos(z_2' + \pi/2) + \cos z_1 \sin(z_1' + \pi/2).$$

Using (6.36) on both sides of the above equation, we obtain

$$\cos(z_1 + z_2') = -\sin z_1 \sin z_2' + \cos z_1 \cos z_1'$$

$$\text{i.e.,} \quad \cos(z_1 + z_2) = \cos z_1 \cos z_2 - \sin z_1 \sin z_2. \tag{6.40}$$

The above equations (6.39) and (6.40) are the exact analogues of the corresponding relations one has for functions of real arguments.

Real and Imaginary Parts

$\sin z$ and $\cos z$ are generally complex quantities. We shall now resolve them into real and imaginary parts:

$$\begin{aligned} 2\cos z &= e^{iz} + e^{-iz} = e^{ix}e^{-y} + e^{-ix}e^{y} \\ &= (\cos x + i \sin x)e^{-y} + (\cos x - i \sin x)e^{y} \\ &= \cos x(e^{y} + e^{-y}) - i \sin x(e^{y} - e^{-y}). \end{aligned}$$

Hence,

$$\cos z = \cos x \cosh y - i \sin x \sinh y. \tag{6.41}$$

Differentiating with respect to z and using $f'(z) = u_x + iv_x$, one obtains

$$\sin z = \sin x \cosh y + i \cos x \sinh y. \tag{6.42}$$

Magnitudes

In particular it follows that

$$|\sin z|^2 = \sin^2 x \cosh^2 y + \cos^2 x \sinh^2 y. \tag{6.43}$$

We use $\cosh^2 y = 1 + \sinh^2 y$ in the above to obtain

$$\begin{aligned}|\sin z|^2 &= \sin^2 x + \sinh^2 y(\sin^2 x + \cos^2 x)\\ &= \sin^2 x + \sinh^2 y.\end{aligned} \tag{6.44}$$

Replacing $x \to x + \pi/2$ in the above and using (6.36) we obtain

$$|\cos z|^2 = \cos^2 x + \sinh^2 y.$$

As $\sinh^2 y$ can take an arbitarily high value for a large enough y ($\sinh^2 y$ grows exponentially with y), there is no absolute upper-bound on $|\sin z|$ and $|\cos z|$ for a complex z. We can, however, establish the inequalities giving lower bounds:

$$|\sin z| \geq |\sin x|; \; |\cos z| \geq |\cos x|.$$

Zeros of $\sin z$ *and* $\cos z$

From Eq. (6.42) it is clear that $\sin z$ vanishes only when

$$\sin x = 0 \quad (\text{as } \cosh y \geq 1 \text{ is never zero})$$

and

$$\cos x \sinh y = 0.$$

When $\sin x = 0$, $\cos x = \pm 1 \neq 0$. Hence second condition requires that $\sinh y = 0$.

Thus, $\sin z = 0$ iff $\sin x = 0$ and $\sinh y = 0$,

$$\begin{aligned}&\text{i.e.,} \quad \text{iff } x = n\pi \quad \text{and} \quad y = 0,\\ &\text{i.e.,} \quad \text{iff } z = n\pi.\end{aligned}$$

Hence, $z = n\pi$ are the only zeros of $\sin z$.

In a similar manner, it is seen that $z = (n + 1/2)\pi$ are the only zeros of $\cos z$.

Other trignometric functions

The other trignometric functions are defined in a manner exactly analogous to the case of a real variable:

$$\tan z = \frac{\sin z}{\cos z}; \quad \cot z = \frac{\cos z}{\sin z} = \frac{1}{\tan z};$$

$$\sec z = \frac{1}{\cos z}; \quad \operatorname{cosec} z = \frac{1}{\sin z}.$$

Their properties can be deduced from those of $\sin z$ and $\cos z$ and are left as exercises.

6.4.3 Logarithmic function

We shall define logarithmic function of a complex variable $z = re^{i\theta}$ in a manner parallel to the case of a real variable. In the case of a real variable the following properties hold: $\log C_1C_2 = \log C_1 + \log C_2$ and $\log e^C = C$. This suggests the following definition for $\log z = \log(re^{i\theta})$

$$\log z = \log r + i\theta \qquad r \neq 0 \tag{6.45}$$

where $\log r$ is the usual natural logarithm (i.e. to the base e) of the real positive number r.

Our aim in this section has been to define functions of a complex variable which

(i) coincide with the corresponding function of a real variable when $z = x > 0$, and
(ii) are analytic in as large a domain as possible. We clearly see that for $z = x > 0, r = x$ and θ can be chosen to be zero. In such a case $\log z = \log r = \log x$. We shall also study analytic properties of this function and establish that the function is indeed analytic in the entire complex plane except for a straight line from origin to $-\infty$; this being the largest domain possible for analyticity of such a function.

6.4.3.1 *Multivalued nature*

The function defined above in (6.45) is a multivalued function. This can be seen as follows. Consider any z. It can always be expressed as

$$z = re^{i\Theta} \qquad -\pi < \Theta \leq \pi.$$

Now using $e^{i(\Theta+2n\pi)} = e^{i\Theta}e^{2\pi in} = e^{i\Theta}$, one sees that z can be written as

$$z = re^{i(\Theta+2n\pi)} \equiv re^{i\theta}.$$

Thus, θ in (6.45) is not unique: one can add to it an integral multiple of 2π without altering z. Hence given z, the imaginary part of $\log z$ is not uniquely defined and $\log z$ is thus a multivalued function. For a given z, $\log z$ has (infinitely) many values.

Principle value of $\log z$

In view of the multivalued nature of $\log z$, it is desirable to define a function for which the range of θ is restricted and which coincides with $\log x$ when $z = x > 0$. We define

$$\operatorname{Log} z = \operatorname{Log} r + i\Theta \qquad -\pi < \Theta \leq \pi.$$

The function is written with a capital L and is called the principle value of $\log z$. Evidently when $z = x > 0$, Θ is zero and $\operatorname{Log} z = \operatorname{Log} r = \operatorname{Log} x$.

As there is exactly one value of Θ for a given z, $\operatorname{Log} z$ is a single-valued function. $\operatorname{Log} z$ is, however, not continuous across the negative y-axis.

6.4.3.2 *Analyticity of* Log z

We want to find the domain of analyticity of Log z. In order that any function $f(z)$ be analytic at $z = z_0$, it is first of all necessary that it is continuous at $z = z_0$. (Otherwise, $f'(z)$ cannot exist there). Now the function is not continuous at any point on the negative x-axis. For just above the −ve real axis, the imaginary part of Log z which is Θ is (nearly) π and just below the −ve real axis, Θ is $-\pi$. Thus Log z suffers a discontinuity across the negative real axis and hence cannot be analytic there.

It is not difficult to show, however, that Log z is analytic everywhere else. First of all, we shall verify CR conditions using polar coordinates.

$$u = \operatorname{Log} r \qquad v = \Theta = \theta \qquad -\pi < \theta < \pi$$
$$r > 0$$

Now, partial derivatives u_r, v_r, u_θ, v_θ exist and

$$u_r = \frac{1}{r};\ v_r = 0;\ u_\theta = 0;\ v_\theta = 1.$$

Hence $u_r = \frac{1}{r}v_\theta$; $\frac{1}{r}u_\theta = -v_r$ are both verified in the entire C plane excluding the −ve real axis and the origin. Moreover, all first ordered partial derivatives are continuous in this domain. Hence Log z is analytic for $r > 0$; $-\pi < \theta < \pi$.

Further, we can find the derivative of Log z as:

$$\frac{d}{dz}\operatorname{Log} z = e^{-i\theta}(u_r + iv_r) = e^{-i\theta}\frac{1}{r} = \frac{1}{z}$$

in this domain.

We shall discuss the Riemann surfaces and branches of log z in Section 8.4.

6.4.4 Complex exponents

The definition we have given for log z (together with that of the exponential) can be used to define complex exponent c of any complex number z. To see how this can be done, consider

$$\log z^n = n \log z : \qquad n \text{ an integer, } z \neq 0,$$

i.e.

$$z^n = \exp(n \log z)$$

valid for an integral exponent n. We generalize the above result into the definition of z^c for any complex c,

$$z^c \equiv \exp(c \log z) \qquad z \neq 0.$$

As $\log z$ is multivalued, $\exp(c \log z)$ is generally multivalued (even though $\exp(n \log z)$ is not multivalued for an integral n. Why?). We may choose a particular value of z^c by choosing a particular values of $\log z$. This is done by choosing the range for θ, viz. $\alpha < \theta < 2\pi + \alpha$ for a real α. Thus,

$$z^c = \exp(c \log z) \qquad |z| > 0 \qquad \alpha < \theta < 2\pi + \alpha$$

is then single-valued in this domain.

Properties: For the function z^c, single-valued in the domain $D\{|z| \neq 0, \alpha < \theta < 2\pi + \alpha\}$,

(1) $z^{c_1} z^{c_2} = z^{c_1+c_2}$, for

$$\begin{aligned} z^{c_1} z^{c_2} &= \exp(c_1 \log z) \exp(c_2 \log z) \\ &= \exp[(c_1 + c_2) \log z] = z^{c_1+c_2} \end{aligned}$$

(2) Inverse: $z^{-c} = \dfrac{1}{z^c}$ since $z^c z^{-c} = z^0 = 1$.

(3) Analyticity and derivative: As $\log z$ is analytic in $\alpha < \theta < 2\pi + \alpha, r > 0$ (for any real α) and the exponential function is entire, z^c will be analytic in the same domain, being an analytic function of an analytic function in this domain. We now proceed to find the derivative.

$$z^c = g[f(z)] \equiv g[w]$$

with $g(w) = e^w$; $w = f(z) = c \log z$. Hence

$$\begin{aligned} \frac{d}{dz} z^c &= \frac{dg}{dw}\frac{dw}{dz} = e^w \frac{c}{z} = g(w)\frac{c}{z} \\ &= \frac{cz^c}{z} \\ &= cz^c z^{-1} = cz^{c-1}. \end{aligned}$$

Thus, $\dfrac{d}{dz} z^c = cz^{c-1} \quad |z| \neq 0; \quad \alpha < \theta < 2\pi + \alpha.$

Problems

On § 6.1

6.1 Prove that if $z_1 z_2 = 0$ at least one of z_1 and z_2 is zero. Generalize the result to the product of three complex numbers.

6.2 Verify that $\frac{1}{z_1 z_2} = \frac{1}{z_1} \cdot \frac{1}{z_2} \qquad (z_1 \neq 0; z_2 \neq 0)$

6.3 Calculate z^{-1} if $z = 5 + 12i$.

6.4 Prove that $|\text{Im } z| + |\text{Re } z| \leq \sqrt{2}|z|$. [Hint: Square both sides and group all terms together.] Give an example of z for which the above relation is an equality.

6.5 Show that $(z^*)^2 = z^2$ if z is either real or purely imaginary.

6.6 Show that for four complex numbers a, b, c and d

$$\frac{|a+b|}{|c-d|} \leq \frac{|a|+|b|}{||c|-|d||}; \qquad |c| \neq |d|$$

6.7 Verify the two triangle inequalities for $z_1 = 5 + 12i$ and $z_2 = 4 - 3i$.

6.8 Show that

(a) $|z_1 + z_2 + z_3| \leq |z_1| + |z_2| + |z_3|$

(b) $|z_1 + z_2 + z_3| \geq |z_1| - |z_2| - |z_3|$

6.9 Starting from the relation Re $(z_1^* z_2) \leq |z_1^* z_2|$, show how both the triangle inequalities can be deduced. Show further that in the above relation, the *equality* is fulfilled when $z_1 = \alpha z_2$ (α real).

On § 6.2

6.10 Find if the following functions are differentiable at $z = 0$ by (i) direct evaluation of $\lim\limits_{z\to 0} \frac{f(z)-f(0)}{z}$ along $y = mx$ (in arbitrary real numbers), (ii) using CR conditions in cartesian coordinates, (ii) using CR conditions in polar coordinates, (iv) using CR conditions with z and z^* as independent variables.

(a) $f(z) = x^2 + ixy$ (b) $f(z) = xy \ln x + iy^2$

(c) $f(z) = \dfrac{1}{x + iy - 1}$

6.11 Show that CR conditions in polar coordinates can be formulated as follows: Express the given function $f(z)$ as the complex function $g(r, \theta)$. Perform a change of variables to $re^{i\theta}$ and θ i.e., $f(z) = g(r, \theta) = h(re^{i\theta}, \theta)$. Then CR conditions for $f(z)$ are also expressible as the complex equation

$$\frac{\partial \text{h}\,(re^{i\theta}, \theta)}{\partial\theta} = 0,$$

where $\frac{\partial}{\partial\theta}$ refers to the partial derivative, keeping the first combination $re^{i\theta}$ fixed.

6.12 Regarding a complex function as $f(x, y)$, show that the CR condition can be expressed as

$$\frac{\partial f}{\partial x} + i\frac{\partial f}{\partial y} = 0.$$

On § 6.3

6.13 Show that the function $f(z) = \frac{1}{z-1}$ is analytic in the disc $|z| < 1$. Verify explicitly the analytic nature at the point $z = 1 - \varepsilon$ (ε small real positive number) by constructing explicitly a neighbourhood where it is differentiable.

6.14 Show that the derivative of a function at an analytic point is given (in polar coordinates) by

$$f'(z) = (u_r + iv_r)e^{-i\theta}.$$

6.15 Show that there is no function analytic in any domain whose real part is $u(x, y) = x^2 + y^2$.

6.16 Show that $u(x, y) = e^{x^2-y^2} \cos xy$ is harmonic everywhere. Find its harmonic conjugate.

6.17 Let $u(x, y)$ be a function of (x, y) (over a domain D) whose derivatives upto second order exist, and u is harmonic in D. Show that there exists a function $f(z)$ analytic in D such that Re $g(z) = u$. Is $f(z)$ unique?

On § 6.4

6.18 Verify the relations (6.34)–(6.37) of §6.4.2.

6.19 Verify, from the defining relation, the property (6.38) of §6.4.2.

6.20 Show that $\sin h(x + \frac{i\pi}{2}) = i \cos hx$.

6.21 Show graphically the region of complex plane for which $|\sin z| > 1$ for $0 \leq x \leq \pi/2$. Do so by considering values of $x = 0, \pi/6, \pi/4, \pi/6, \pi/2$.

6.22 Find all roots of the equation $\sin z = 3$.

6.23 Show that generally $|\cos z| \geq \sqrt{|1 - |\sin z|^2|}$ and that the equality is fulfilled only when z is real.

6.24 Show that $|e^{f(z)}| \leq e^{|f(z)|}$, where $f(z)$ is any function of z.

6.25 Find the region of the complex plane for which

$$0 \leq \text{Im}\, e^{z^2} \leq 1.$$

6.26 Solve the equation $\sin z = 1 + i/2$. Find $\cos z$ and $\tan z$. Also find $\sin hz$.

6.27 Find a solution of $\sin z = \sin hz$, $(z \neq 0)$.

6.28 Find all zeros of $\sin hz$.

6.29 Show that $|\sin hz|^2 \leq \cos h^2 x$ for any y.

6.30 Find values of z for which Log $\sin z$ is a real positive number.

6.31 Show that $\text{Log}\, i^5 \neq 5\,\text{Log}\, i$.

6.32 Discuss the behaviour of $|e^{1/z}|$ as z approaches zero along (i) +ve real axis (ii) −ve real axis (iii) imaginary axis.

7 Complex Variables – II

In this chapter, we shall first introduce the contour integrals of a function of a complex variable z, viz. $\int_C f(z)dz$, over a contour C. We shall, with the use of these contour integrals, establish a fundamental result in the theory of analytic functions, the Cauchy–Goursat Theorem. The remainder of the chapter will deal with a number of results arising out of this result; they will involve the Cauchy Integral formula, the Residue Theorem and a large number of results that can be established with the help of these. We will also derive the Taylor and Laurent expansions for analytic functions around its regular or singular points. We will then study general properties of power series. Finally, we shall discuss the applications of the Residue Theorem towards evaluation of real definite integrals.

7.1 Contour Integrals

Consider a curve C from a to b in the complex plane as shown in Fig. 7.1. The curve can be parameterized by a real parameter t as follows:

$$x = x(t), \quad y = y(t) \quad : \quad t_a \leq t \leq t_b \tag{7.1}$$

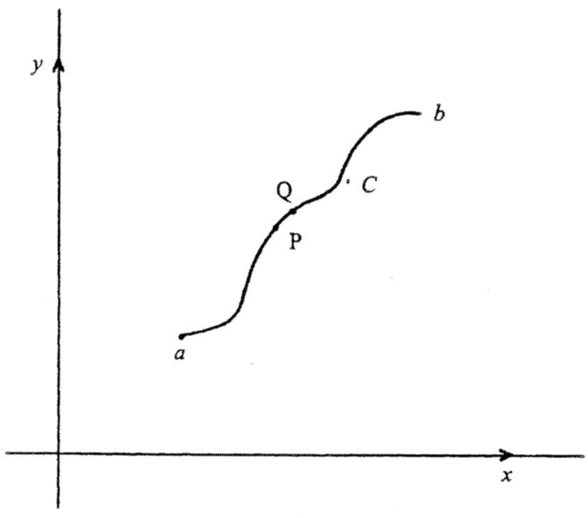

Fig. 7.1 *A continuous curve ab*

(Here, t_a and t_b are the values of the real parameter at the end points.) We assume that t increases monotonically as the curve is traversed from a to b, so that $x(t)$ and $y(t)$ are real single-valued functions of t.

Definitions:

(a) The curve C is an arc if $x(t)$ and $y(t)$ are continuous functions of t.
(b) An arc that does not cross itself is a simple arc. In this case every pair (x, y) on the curve has only one t associated with it.
(c) An arc is regular (smooth) if $x'(t)$ and $y'(t)$ exist and are continuous with $(x'^2(t) + y'^2(t)) \neq 0$ everywhere.
(d) A contour is a piecewise regular arc, i.e., a continuous curve consisting of a finite number of regular arcs.
(e) A closed contour that does not cross itself is a simple contour.

7.1.1 Definition of a contour integral

Consider a function $f(z)$ of a complex variable defined over a region which contains a contour ab such as that shown in Fig. 7.1. We want to define $\int_C f(z)dz$ for this function over this contour. We wish to do this using our knowledge of the Riemann integral over a real variable. We note that on the curve C, z can be expressed as a function of t in accordance with (7.1), viz.,

$$z = z(t) \quad \text{on } C. \tag{7.2}$$

Thus, *on* C, $f(z)$ can be expressed as a function of t

$$f(z) = f[z(t)] \quad \text{on } C. \tag{7.3}$$

On C, we may take two neighbouring points P and Q represented by z and $z + \Delta z$. Then,

$$\Delta z = \Delta x + i\Delta y = \Delta x(t) + i\Delta y(t).$$

In the limit $\Delta z \to 0$, $\Delta t \to 0$ (as $|\frac{dz}{dt}| = \sqrt{x'^2(t) + y'^2(t)}$) exists and is nonzero)*. We can then express, in the limit $\Delta z \to 0$,

$$dz = \frac{dx(t)}{dt}dt + i\frac{dy(t)}{dt}dt. \tag{7.4}$$

On the other hand, $f(z)$ can be expressed in terms of its real and imaginary parts as functions of t.

$$f(z) = u + iv = u[x(t), y(t)] + iv[x(t), y(t)] \equiv u(t) + iv(t). \tag{7.5}$$

*The countour C under consideration may be a collection of more than one smooth arc. In that case $|\frac{dz}{dt}|$ may not exist at the endpoints of each arc. Such points, finite in number, are being excluded in the definition below.

Then,

$$f(z)dz = [u(t)x'(t) - v(t)y'(t)]dt + i[v(t)x'(t) + u(t)y'(t)]dt. \qquad (7.6)$$

Now the ordinary Riemann integral of each the real part and the imaginary part above is known. We then define the contour integral of $f(z)$ over the contour C as the sum of the Riemann integrals over t:

$$\int_C f(z)dz = \int_{t_a}^{t_b} (ux' - vy')dt + i \int_{t_a}^{t_b} (vx' + uy')dt. \qquad (7.7)$$

As a line integral over the x–y plane, the above is easily seen to be

$$\int_C f(z)dz = \int_C (udx - vdy) + i \int_C (vdx + udy). \qquad (7.8)$$

Generally the line integrals on the R.H.S of (7.8) depend on C, the path, and hence the contour integral of a function is (generally) dependent on the choice of C and not just the end points a and b.

7.1.2 Darboux inequality

In the future discussions, we shall very often have occasion to put an upper bound on the contour integrals of the type just introduced.

Consider the contour integral

$$\int_C f(z)dz = \int_{t_a}^{t_b} f(z)\frac{dz}{dt}dt$$

where the R.H.S is understood as (a sum of two) Riemann integrals with the complex integrand $g(t) \equiv f[z(t)]\frac{dz}{dt}$. Hence following the definition of the Riemann integral.

$$I = \int_{t_a}^{t_b} f(z)\frac{dz}{dt}dt = \lim_{\Delta t_k \to o} \sum_{k=o}^{n} g(t_k)(t_{k+1} - t_k) \qquad t_0 \equiv t_a,\ t_n = t_b.$$

Now

$$\left| \sum_{k=0}^{n} g(t_k)(t_{k+1} - t_k) \right| \le \sum_{k=0}^{n} |g(t_k)|\Delta t_k$$

$$= \sum_{k=0}^{n} |f[z(t_k)]| \left| \frac{dz}{dt}(t_k) \right| \Delta t_k.$$

Now *suppose* there exists an upper bound for $|f(z)|$ over the contour C

$$|f[z(t)]| \le M \qquad t_a \le t \le t_b.$$

Then *

$$|I_n| \equiv \left|\sum_{k=0}^{n} g(t_k)\Delta t_k\right| \leq M\sum_{k=0}^{n}\left|\frac{dz}{dt}(t_k)\right|\Delta t_k \leq M\sum_{k=0}^{n}|dz_k| \leq ML$$

here.

Now as I is the limit of the sequence (assumed to exist), given ε however small, there exists N such that for $n > N$,

$$|I - I_n| < \varepsilon$$

$$\text{i.e.,}\quad |I| < |I_n| + \varepsilon \leq ML + \varepsilon. \qquad n \geq N$$

As I is independent of n, we therefore have

$$|I| \leq ML.$$

7.2 Cauchy Theorem

Now we shall proceed to establish, by far the most powerful result on the analytic functions in this book, and a very basic one which serves as a starting point for a number of subsequent results in this chapter. The result in the form given a little later was proved by Cauchy. Its proof is based upon a theorem involving line integrals of real functions of two real variables, viz. the Green's theorem, which we shall proceed to discuss shortly. We shall only *state* the Green's theorem without proof; the latter can however be given along the lines of the proof for Stokes' theorem.

7.2.1 Green's theorem

Let two real valued functions $f(x, y)$ and $g(x, y)$ together with their first order partial derivatives be continuous throughout a closed region R consisting of point interior to and on a simple closed contour C in the x–y plane. C is assumed to be described in the positive (anticlockwise) sense. Then,

$$\oint_C [f(x, y)dx + g(x, y)dy] = \int\int \left(\frac{\partial g}{\partial x} - \frac{\partial f}{\partial y}\right) \quad dxdy.$$

The integral on the R.H.S. is taken over the area enclosed by C.

*In the argument below we have replaced $|\frac{dz}{dt}|\Delta t_k$ by $|dz_k|$. This is so, since we are going to let $\Delta t_k \to 0$. To make it more rigorous, we should replace $\sum_k |\frac{dz}{dt}|\Delta t_k$ by $\sum_k |dz_k| + nO(\Delta t)^2$. The extra term vanishes as $\Delta t \to 0$.

Example We shall verify Green's theorem for the following simple example.

Consider $f(x, y) = y^2$ and $g(x, y) = x$. Consider the contour to be a triangle as shown in Fig 7.2. Clearly f and g satisfy the conditions of the Green's theorem. Consider the left-hand side of the Green's theorem as applicable to this example:

$$\oint_C \left[y^2 dx + x dy\right] \equiv \int_{\text{OA}} + \int_{\text{AB}} + \int_{\text{BO}}$$

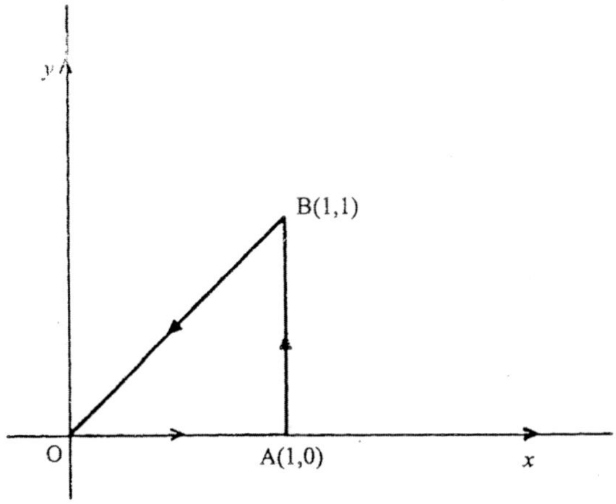

Fig. 7.2 *Contour OABO for application of Green's theorem*

On OA, $dy = 0$; on AB, $dx = 0$ and on OA, $x = y$ so that $dx = dy$. Hence,

$$\int_{\text{OA}} \left[y^2 dx + x dy\right] = \int_{\text{OA}} y^2 dx = \int 0 dx = 0$$

$$\int_{\text{AB}} \left[y^2 dx + x dy\right] = \int_{\text{OA}}^{1} x dy = \int_0^1 1.dy = 1$$

$$\int_{\text{BO}} \left[y^2 dx + x dy\right] = \int_1^0 \left[y^2 dx + x dx\right] = \int_1^0 (x^2 + x) dx = -\frac{5}{6}$$

Hence $$\oint_C \left[y^2 dx + x dy\right] = \frac{1}{6}.$$

Now consider the R.H.S of the Green's theorem.

$$\iint \left(\frac{\partial g}{\partial x} - \frac{\partial f}{\partial y} \right) dxdy = \iint (1-2y) \quad dxdy$$
$$= \int_0^1 dy(1-2y) \int_y^1 dx$$
$$= \int_0^1 dy(1-2y)(1-y) = \frac{1}{6}.$$

This verifies the Green's theorem in this example.

7.2.2 The Cauchy theorem

Theorem 7.1 Let f be analytic in a closed region R and f' be continuous in R. Let C be a simple closed contour in R. Then

$$\oint_C f(z)dz = 0.$$

Proof: Firstly, we recall,

$$f'(z) = u_x + iv_x = -iu_y + v_y.$$

Now, since $f'(z)$ in continuous in R so are its real and imaginary parts. Hence all first order partial derivatives u_x, u_y, v_x, v_y are continuous in R. Now consider

$$\oint_C f(z)dz = \oint_C [u(x,y)dx - v(x,y)dy] + i \oint_C [v(x,y)dx + u(x,y)dy].$$

Now, we shall apply the Green's theorem to each of these integrals. We note that analyticity of f and continuity of $f'(z)$ in R assure the continuity of u, v and of partial derivatives u_x, u_y, v_x and v_y in R; thus the conditions of Green's theorem are fulfilled. We then have

$$\oint_C f(z)dz = \int\int (-v_x - u_y)dxdy + i \int\int (u_x - v_y)dxdy.$$

But the R.H.S vanishes on account of CR conditions $u_x = v_y$ and $v_x = -u_y$ which are obeyed by u and v since f is analytic in R. This completes the proof.

Examples Here we present two examples, one of which illustrates the Cauchy theorem and another illustrates what happens when conditions of Cauchy's theorem are not fulfilled.

(1) Consider $f(z) = z^2$ and let C be the unit circle $|z| = 1$. Then

$$\oint_C f(z)dz = \oint z^2 dz \qquad \text{over the unit circle.}$$

We expect this to vanish as $f(z)$ is entire and, in particular it is analytic on and within C and $f'(z) = 2z$ is continuous in this region. We verify this by direct evaluation. On a unit circle,

$$z = e^{i\theta}, \qquad 0 \leq \theta < 2\pi$$

$$dz = ie^{i\theta}d\theta,$$

$$\oint z^2 dz = \int_0^{2\pi} e^{2i\theta} ie^{i\theta} d\theta = i\int_0^{2\pi} e^{3i\theta} d\theta = \frac{i}{3i}\left[e^{3i\theta}\right]_0^{2\pi} = 0.$$

Thus, Cauchy's theorem is verified. This result applies as well to $f(z) = z^n$; n a positive integer.

(2) Consider $f(z) = z^{1/2}$ and again let C be the unit circle. Here $f(z)$ is *not* analytic within the unit circle. It has a branch cut* which may be taken along the negative real axis. Then, it is, in fact, not even continuous across the negative real axis. Hence the conditions of Cauchy's theorem are not met and we cannot expect the integral to vanish. As seen below a straightforward evaluation in fact shows that the integral does not vanish.

$$\oint z^{1/2} dz = \int_{-\pi}^{\pi} e^{i\theta/2} ie^{i\theta} d\theta = i\int_{-\pi}^{\pi} e^{3/2i\theta} d\theta$$

$$= \frac{2}{3}\left[e^{3/2i\theta}\right]_{-\pi}^{\pi} = \frac{2}{3}\left[e^{3/2\pi i} - e^{-3/2\pi i}\right]$$

$$= -\frac{4i}{3} \neq 0.$$

7.2.3 The Cauchy–Goursat theorem

It turns out that the conditions for the validity of the Cauchy's result as stated originally by Cauchy are not the minimum necessary ones. Goursat showed that the condition that "$f'(z)$ be continuous within and on C" is redundant and that the same result can be proved without it. This result, with weaker conditions on $f(z)$, is referred to as the Cauchy–Goursat theorem which states :

Theorem 7.2 If a function f is analytic within and on a simple closed contour C, then

$$\oint_C f(z)dz = 0.$$

We shall not present the proof in this book. We shall henceforth use the theorem in this less restrictive form however.

*For definition of a branch cut see Section 7.8

7.2.4 Generalizations

Generalization of the Cauchy–Goursat theorem for multiply-connected domains and for contours that are not simple is made in a straightforward manner. These generalizations are best seen with the help of simple examples.

Example (1) Consider the contour C shown in Fig. 7.3 (a) in the domain of analyticity of $f(z)$. The contour C is traversed as shown by arrows along ABDAEFA. We note that we can regard C as consisting of two parts $C_1 \equiv$ AEFA going anticlockwise and $C_2 \equiv$ ABDA going clockwise.

Now, if $f(z)$ is analytic on and within C, it is so on C_1, C_2 and within C_1 and C_2. Thus we have

$$\oint_{C_1} f(z)dz = 0; \qquad \oint_{-C_2} f(z)dz = 0 = -\oint_{C_2} f(z)dz \tag{7.9}$$

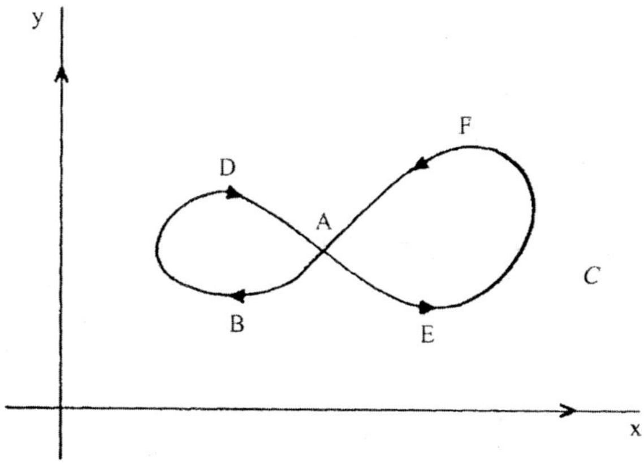

Fig. 7.3(a) *Cauchy–Goursat theorem: non-simple contour*

(where $-C_2$ runs anticlockwise). Thus,

$$\oint_C f(z)dz = \left[\int_{C_1} + \int_{C_2}\right] f(z)dz = 0.$$

Thus the Cauchy theorem is established for the closed contour C which is not simple.

Example (2) Consider the contour C in the Fig. 7.3 (b) that constitutes the boundary of a multiply-connected domain. C consists of two disconnected parts: C_1 traversed in the anticlockwise sense, and C_2 (enclosed by C_1) traversed in the clockwise sense. Let f be analytic on and within C (i.e., in the closed region enclosed between C_1 and C_2). We can then show the Cauchy–Goursat theorem for C in the following form:

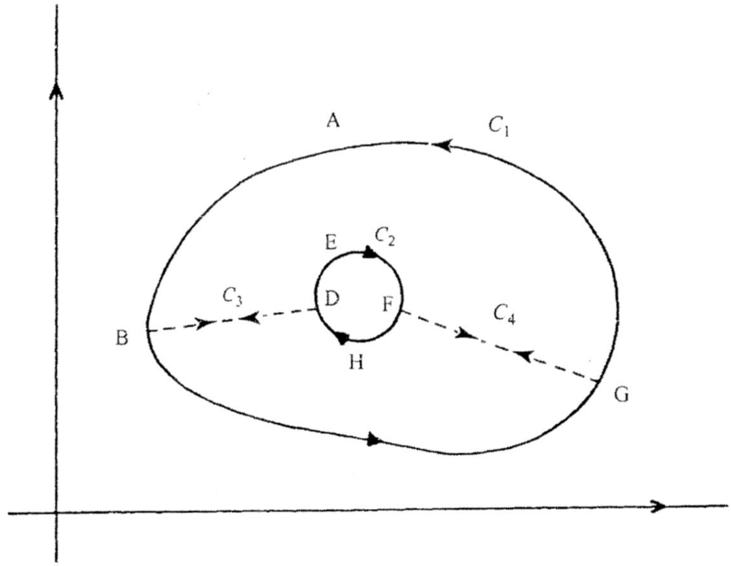

Fig. 7.3(b) *Cauchy–Goursat theorem: multiply-connected domain*

$$\oint_C f(z)dz \equiv \oint_{C_1} f(z)dz + \oint_{C_2} f(z)dz = 0 \tag{7.10}$$

from the original result for simple closed curves. To see this, we introduce bridging contours C_3 and C_4 as shown. We now apply the Cauchy–Goursat theorem to the simply-connected contours $C_5 \equiv$ ABDEFGA and $C_6 \equiv$ BGFHDB.

$$\oint_{C_5} f(z)dz = 0 = \oint_{C_6} f(z)dz.$$

Adding these results and noting that contributions over BD and FG cancel out in the sum, one recovers (7.10)

7.3 Cauchy's Integral Formula

We now proceed to obtain one important application of the Cauchy–Goursat theorem: the Cauchy Integral formula. The Cauchy integral formula shows that the values of an analytic function f in the interior of a (closed) region D are entirely determined by those on the boundary of the region D. (Here f is analytic within D *and* on the boundary). A further extension of the methods evolved in proving Cauchy's integral formula will enable one to determine arbitrary order derivatives of the function $f(z)$ (*shown* to exist) in the interior of D in terms of only the boundary values of the function $f(z)$. The result on the Cauchy integral formula states:

Theorem 7.3 Let a function $f(z)$ be analytic in and on a simple closed contour C (taken in the anticlockwise sense). Then the value of $f(z)$ at any point z_0 interior to C is given by the contour integral

$$f(z_0) = \frac{1}{2\pi i} \oint_C \frac{f(z)dz}{z - z_0}.$$

Proof: Enclose the point z_0 (in the interior of C) by a sufficiently small circle C_0 of radius ε which is in the interior of C as shown in the Fig. 7.4. We take the circle C_0 in the clockwise sense. Consider the (multiply-connected) domain D_1 bounded by C and C_0. In this domain $D_1 \subset D$, $f(z)$ is analytic. $\frac{1}{z-z_0}$ is also analytic in D_1 (as the only singular point $z = z_0$ has been excluded). Hence $\frac{f(z)}{z-z_0}$ is analytic in D_1 and also on the boundary (C together with C_0). Hence by Cauchy–Goursat theorem applied to D_1 will yield

$$\oint_C \frac{f(z)}{z - z_0} dz + \oint_{C_0} \frac{f(z)}{z - z_0} dz = 0. \tag{7.11}$$

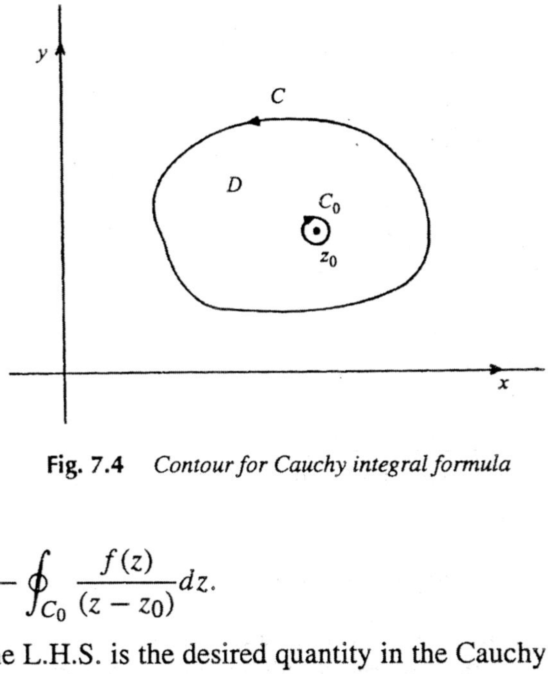

Fig. 7.4 *Contour for Cauchy integral formula*

Hence

$$\oint_C \frac{f(z)}{z - z_0} dz = -\oint_{C_0} \frac{f(z)}{(z - z_0)} dz. \tag{7.12}$$

Now the integral on the L.H.S. is the desired quantity in the Cauchy integral formula and is clearly independent of ε, the radius of C_0. Hence, the integral on the R.H.S. must in fact be independent of ε. In particular, it can be evaluated by letting ε, the radius of C_0, go to zero. As $f(z)$ is continuous, $f(z)$ will approach $f(z_0)$ on C_0 and the integral on R.H.S will approach

$$-f(z_0) \oint_{C_0} \frac{dz}{z - z_0} = 2\pi i f(z_0). \tag{7.13}$$

(See Problem 7.3.) To exhibit this more rigorously, consider

$$\left|\oint_{C_0} \frac{[f(z)-f(z_0)]}{z-z_0}dz\right| \leq \left[\frac{|f(z)-f(z_0)|}{|z-z_0|}\right]_{\max} 2\pi\varepsilon$$

Now on C, $|z-z_0| = \varepsilon$, and as $f(z)$ is analytic and hence continuous for a given δ, however small, we can find ε such that

$$|f(z)-f(z_0)| < \delta \qquad \text{for} \quad |z-z_0| \leq \varepsilon.$$

Hence

$$\left|\oint_{C_0} \frac{f(z)-f(z_0)}{z-z_0}dz\right| \leq \frac{\delta}{\varepsilon}2\pi\varepsilon \leq 2\pi\delta. \tag{7.14}$$

And as this integral on the L.H.S. of (7.14) is independent of ε and according to (7.14), it can be made arbitrarily small by choosing ε, it must, in fact, be zero. This proves that the R.H.S. of (7.12) and L.H.S. of (7.13) are in fact equal. Equations (7.12) and (7.13) then prove the Cauchy integral formula.

Example Let C be the unit circle $|z| = 1$ (positively oriented). Consider $f(z) = \frac{z^2}{\left[z+\frac{7}{2}\right]\left[z-\frac{5}{2}\right]}$. $f(z)$ is analytic on and within C. Let $z_0 = \frac{1}{2}$. Then the Cauchy Integral formula implies

$$\oint_C \frac{f(z)dz}{z-z_0} = \oint_C \frac{z^2 dz}{\left[z-\frac{1}{2}\right]\left[z+\frac{7}{2}\right]\left[z-\frac{5}{2}\right]} = 2\pi i f\left(\frac{1}{2}\right)$$

$$= 2\pi i \frac{(1/4)}{4(-2)} = -\frac{\pi i}{16}.$$

This result can be generalized to a multiply-connected region. Then the contour integral in the statement of Theorem 7.2 is replaced by a sum of contour intervals over the boundaries taken in an appropriate sense. See for e.g., discussion at the end of Section 7.2.4.

7.4 Derivatives of an Analytic Function

The Cauchy's integral formula shows that if the values of an analytic function $f(z)$ on a simple closed contour C (within the region of analyticity) are given, $f(z)$ is determined at every point within C. It thus follows that as $f'(z_0)$ is determined uniquely by values of $f(z)$ in a neighborhood of z_0, it should be possible to express $f'(z_0)$ also entirely in terms of values of $f(z)$ on C. This argument can then be iterated successively to conclude that $f^{(n)}(z_0)$ for any n can be determined entirely in terms of values of $f(z)$ on C. To see how this could work, consider again the Cauchy integral formula

$$f(z_0) = \frac{1}{2\pi i}\oint_C \frac{f(z)dz}{z - z_0}. \tag{7.15}$$

We re-express this by taking z as the point in the interior and w as the variable of integration

$$f(z) = \frac{1}{2\pi i}\oint_C \frac{f(w)dw}{w - z}. \tag{7.16}$$

Now suppose we could carry out differentiation of the R.H.S. with z *inside* the integral. The only z-dependence is in the factor $\frac{1}{w-z}$ (recall that C is fixed). Hence, we would have

$$f'(z) = \frac{1}{2\pi i}\oint_C \frac{f(w)dw}{(w - z)^2} \tag{7.17}$$

and if we were to carry out differentiations successively we would have

$$f^{(n)}(z) = \frac{n!}{2\pi i}\oint_C \frac{f(w)dw}{(w - z)^{n+1}}. \tag{7.18}$$

Of course, the validity of interchange of differentiation and integration is not obvious. We shall prove it for (7.17), leaving (7.18) as an exercise.

Before we proceed with the proof, we note the accompanying details in Fig. 7.5. z is a point P in the interior of C. Let Q be the point on C nearest to P. Let $PQ = d > 0$. Then $|w - z| \geq d$ for every w on C. Now we note from (7.16)

$$\frac{f(z+\Delta z) - f(z)}{\Delta z} = \frac{1}{2\pi i \Delta z}\oint_C dw f(w)\left[\frac{1}{w - z - \Delta z} - \frac{1}{w - z}\right]$$

$$= \frac{1}{2\pi i}\oint_C \frac{dw f(w)}{(w - z - \Delta z)(w - z)}. \tag{7.19}$$

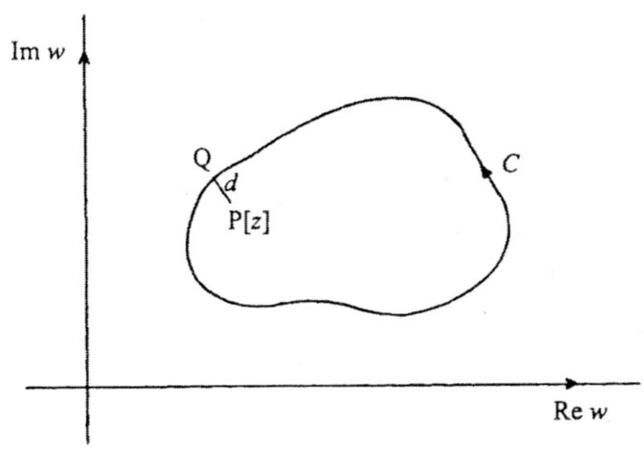

Fig. 7.5

(Here Δz has been chosen small enough so that $z+\Delta z$ is also within C so that $|\Delta z| < d$). We want to show that the R.H.S. of (7.19) approaches the R.H.S. of (7.17) as $\Delta z \to 0$. Hence, we consider the difference:

$$\Delta I = \frac{1}{2\pi i}\oint_C dw f(w)\left[\frac{1}{(w-z-\Delta z)(w-z)} - \frac{1}{(w-z)^2}\right]$$

$$= \frac{\Delta z}{2\pi i}\oint_C \frac{dw f(w)}{(w-z-\Delta z)(w-z)^2}. \tag{7.20}$$

This, evidently, would go to zero as $\Delta z \to 0$, provided of course the contour integral remains bounded in this limit. This happens essentially because

(i) $f(w)$, being analytic on C, is bounded on C: $|f(w)| < M$ for some M on C;
(ii) length of the contour L is finite;
(iii) the denominators are bounded below in magnitude as $|w-z| \geqslant d$; $|w-z-\Delta z| \geq |w-z| - |\Delta z| \geqslant d - |\Delta z| > 0$. Thus,

$$\left|\oint_C \frac{dw f(w)}{(w-z-\Delta z)(w-z)^2}\right| < \frac{ML}{d^2(d-|\Delta z|)}. \tag{7.21}$$

Thus as $\Delta z \to 0$, ΔI of (7.20) goes to zero. This in turn implies

$$\lim_{\Delta z\to 0}\frac{f(z+\Delta z)-f(z)}{\Delta z} = f'(z) = \frac{1}{2\pi i}\oint_C \frac{dw f(w)}{(w-z)^2}$$

Equation. (7.18) can then be proved by induction. Assume the result (7.18) for $n = m-1$ and evaluate $f^{(m)}(z)$ as in (7.19). Analogous manipulations following (7.19), somewhat modified in detail, then allow one to arrive at (7.18) for $n = m$. As the result is proven for $m = 1$ (the Cauchy integral formula itself), (7.18) is then established (see Problem 7.12).

Example Consider the function $f(z) = \frac{1}{1-z}$ and let C be the positively oriented circle $|z| = \frac{1}{2}$. Then $f(z)$ is analytic within and on C. Then, using the Cauchy integral formula, one has

$$f(0) = 1 = \frac{1}{2\pi i}\oint_C \frac{dz}{(1-z)z}$$

and using $f^{(n)}(0) = n!$, we obtain from (7.18)

$$f^{(n)}(0) = n! = \frac{n!}{2\pi i}\oint_C \frac{dz}{(1-z)z^{n+1}}$$

or

$$\oint_C \frac{dz}{(1-z)z^{n+1}} = 2\pi i \qquad n = 0, 1, 2, \ldots.$$

Corollary: Let $f(z)$ be analytic within and on a simple closed curve C. Then derivatives of $f(z)$ to all orders are analytic functions within C.

This is seen as follows. We have seen that $f''(z)$ exists in the interior of C and is given by (7.18) with ($n = 2$). Thus consider any point z in the interior of C. Then there exists a neighbourhood of z (entirely within the interior of C) in which the derivative of the function $f'(z)$ (viz. $f''(z)$) exits. Hence, $f'(z)$ is analytic at z by definition. This argument can further be extended to all n.

7.5 Taylor Series

The reader must be familiar with the Taylor expansion of a function of one variable. To begin with, we will first try to understand what is novel about the Taylor expansion of an analytic function of z as compared with that of a bivariate function. Consider a function $f(z)$ analytic in D that encloses the point $z = z_0$. Let $f(z) = u(x, y) + iv(x, y)$. Now, u and v are functions of two variables x and y. As the reader knows that for the Taylor expansion of function of one variable, (valid under appropriate conditions):

$$g(x) = g(x_0) + (x - x_0)g'(x_0) + \frac{1}{2}(x - x_0)^2 g''(x_0) + \ldots$$

This result can be easily generalized to a function $u(x, y)$ of two variables. Thus under appropriate conditions, on the partial derivatives of $u(x, y)$ a Taylor expansion for u in powers of $(x - x_0)$ and $(y - y_0)$ would exist around $x = x_0$ and $y = y_0$. A similar expansion for $v(x, y)$ could also be written down. Combining these an appropriate expansion for $f(z)$ around $z = z_0$ in powers of $(x - x_0)$ and $(y - y_0)$ would be obtained. (This discussion would apply to any sum $u(x, y) + iv(x, y)$—not just an analytic function—provided certain conditions on partial derivatives of u and v are met). When however $[u(x, y) + iv(x, y)]$ represents a function *analytic* in D, we shall see that (see Problem 7.18):

(a) The conditions on the partial derivatives of u and v necessary for the existence of Taylor expansion are always met in D.

(b) The entire series can be written as a series in the *single complex* variable $(z - z_0)$ instead of *two* real variables $(x - x_0)$ and $(y - y_0)$. This requires certain relations between the partial derivatives of u and v.

(c) The coefficient of $(z - z_0)^n$ in the above series can be written entirely in terms of $f^{(n)}(z_0)$ (rather than nth degree partial derivatives of u and v which can be, *a priori*, very many in number).

As will be seen in Problem 7.19, (a) is a consequence of analyticity of $f(z)$ ensuring existence of $f^{(n)}(z)$. (b) and (c) above are consequences of the CR conditions.

Now we shall proceed to prove the Taylor's theorem for an analytic function of a complex variable.

Theorem 7.4 Let a function $f(z)$ be analytic within a circle C of radius R and with center at $z = z_0$. Then it can be expanded as a series at any point z within C as

$$f(z) = f(z_0) + (z - z_0)f'(z_0) + \frac{(z - z_0)^2}{2!}f''(z_0) + \dots + \frac{f^{(n)}(z_o)}{n!}(z - z_0)^n + \dots \tag{7.22}$$

Qualitative discussion regarding the proof

1. For any function analytic on and within a contour C_1 (to be soon introduced) which encloses z, the Cauchy integral formula expresses $f(z)$ as

$$f(z) = \frac{1}{2\pi i}\int_{C_1} \frac{f(\xi)d\xi}{\xi - z} \tag{7.23}$$

(where ξ is the complex *variable* and z is *fixed*).

2. We note that the integrand on the R.H.S has a simple dependence on z, viz. through the function

$$\frac{1}{\xi - z} \equiv \frac{1}{(\xi - z_0) - (z - z_0)} \equiv \frac{1}{\zeta - Z}, \tag{7.24}$$

irrespective of f and whose expansion in powers of $\frac{Z}{\zeta}$ is well known.

$$\frac{1}{\zeta - Z} = \frac{1}{\zeta}\left[1 + \left(\frac{Z}{\zeta}\right) + \left(\frac{Z}{\zeta}\right)^2 + \cdots\right]; |Z| < |\zeta|. \tag{7.25}$$

(We have not yet discussed the convergence of an infinite complex series; so in fact we will not be using (7.25) directly. However, we note that the above series is in fact a prototype Taylor series that leads to the general Taylor expansion.)

3. When (7.25) is substituted within the integrand of (7.23) (and if we could interchange the sum and the integrand), we would obtain an infinite series in powers of $z = z - z_0$ with coefficients independent of z and which can in fact be identified with the integral representations for $f^{(n)}(z_0)$, $n = 0, 1, \dots$ developed in the previous section.
4. Thus remains only the subtle mathematical point in (3) above regarding the interchange of the summation in (7.25) and the integration. This is taken care of by substituting the *finite* series

$$\frac{1}{\zeta - Z} = \frac{1}{\zeta}\left[1 + \frac{Z}{\zeta} + \left(\frac{Z}{\zeta}\right)^2 + \dots + \left(\frac{Z}{\zeta}\right)^{n-1} + \frac{(Z/\zeta)^n}{1 - Z/\zeta}\right] \tag{7.26}$$

(verified easily in a direct fashion for any n) instead of (7.25) in the integrand of (7.23). Then the sum and the integral can be interchanged. It then, however, remains to be shown that the contribution to (7.23) from the last term in (7.26) is indeed vanishing as $n \to \infty$.

This outlines the proof to which we now proceed.

Figure 7.6 shows the circular contour C_1 (with center at z_0) taken within C but so as to enclose the point z. This is always possible as z is within C. Then for any point ξ on C_1, $|z - z_0| < |\xi - z_0| < R$.

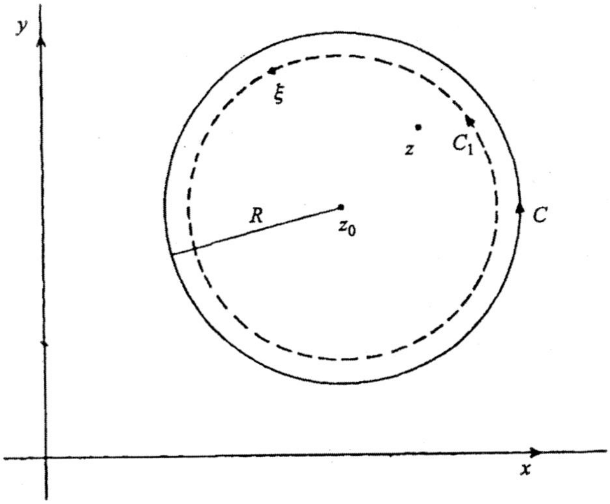

Fig. 7.6

We now, note that $f(z)$ can be expressed by the Cauchy integral formula (7.23) as $f(z)$ is analytic on and within C_1. In (7.23) we then substitute (7.26) to obtain

$$2\pi i f(z) = \oint_{C_1} \frac{d\xi f(\xi)}{\zeta}\left[1 + \frac{Z}{\zeta} + \left(\frac{Z}{\zeta}\right)^2 + \cdots + \left(\frac{Z}{\zeta}\right)^{n-1} + \frac{(Z/\zeta)^n}{1 - Z/\zeta}\right]$$

$$= \oint_{C_1} \frac{d\xi f(\xi)}{\xi - z_0} + Z\oint_{C_1} \frac{d\xi f(\xi)}{(\xi - z_0)^2} + \cdots Z^{n-1}\oint_{C_1} \frac{d\zeta f(\xi)}{(\xi - z_0)^n} + R_n, \quad (7.27)$$

where R_n is the residue after n terms in the above series, viz.

$$R_n \equiv Z^n \oint_{C_1} \frac{f(\xi)d\xi}{(\xi - z_0)^n(\xi - z)}. \quad (7.28)$$

We now note that the coefficients of various powers of Z are identified from Section 7.4 as being equal to

$$f(z_0),\ f'(z_0),\ \ldots\ldots \frac{1}{(n-1)!} f^{(n-1)}(z_0).$$

Then (7.27) reads

$$f(z) = f(z_0) + Zf'(z_0) + \frac{Z^2}{2!} f''(z_0) + \cdots + \frac{Z^{n-1}}{(n-1)!} f^{(n-1)} z_0 + \frac{R_n}{2\pi i}. \tag{7.29}$$

It only remains now to show that $R_n \to 0$ as $n \to \infty$. This will then establish (7.22). This is done by the now familiar method of putting an upper bound on R_n and showing that this vanishes as $n \to \infty$. Note

$$|R_n| = |Z|^n \left| \oint_{C_1} \frac{f(\xi) d(\xi)}{(\xi - z_0)^n (\xi - z)} \right|. \tag{7.30}$$

Now as $f(\xi)$ is analytic on C_1, there exists M such that $|f(\xi)| < M$ for ξ on C_1.
Also,

$$|Z| = |z - z_0| < |\xi - z_0| = R_1;$$

$$|\xi - z| \geq |\xi - z_0| - |z - z_0| = R_1 - |Z|.$$

Thus the integrand in (7.30) is bounded as

$$\left| \frac{f(\xi)}{(\xi - z_0)^n (\xi - z)} \right| \leq \frac{M}{R_1^n (R_1 - |Z|)}.$$

Thus,

$$|R_n| \leq |Z|^n \frac{M}{R_1^n (R_1 - |Z|)} = \frac{M}{R_1 - |Z|} \left| \frac{Z}{R_1} \right|^n$$

$$\to 0 \text{ as } n \to \infty \text{ since } |Z| < R_1.$$

This establishes the Taylor series or the Taylor expansion (7.22) of $f(z)$ around $z = z_0$.

Definition: The Taylor series for $f(z)$ (analytic at $z = 0$) around $z = 0$ is also called the Maclaurin Series.

Examples

(i) Let $f(z) = e^z$ and $z_0 = 0$. Then $f'(z) = e^z = f(z)$ and hence $f^{(n)}(z) = e^z = f(z)$. Thus, $f^{(n)}(z_0) = 1, n = 0, 1, 2, \ldots$. We thus have a Taylor expansion for e^z as

$$e^z = 1 + z + \frac{z^2}{2!} + \cdots + \frac{z^n}{n!} + \cdots$$

As $f(z)$ is entire, the expansion holds for all z.

(ii) Let $f(z) = \frac{1}{1-z}$. Then it is easily seen that

$$f^{(n)}(z) = \frac{n!}{(1-z)^{n+1}}.$$

(a) We could now write the Taylor series about $z_0 = 0$. In that case $f^{(n)}(0) = n!$ $(n = 0, 1, 2, \ldots)$
Then one has a Taylor expansion

$$f(z) = \frac{1}{1-z} = 1 + z + z^2 + \cdots + z^n + \cdots = \sum_{n=0}^{\infty} z^n.$$

As $f(z)$ is analytic within the unit circle, the above expansion holds for $|z| < 1$. (Also, as we shall see, it cannot hold at any point outside the unit circle.)

(b) We could also carry out the expansion about $z_0 = \frac{1}{2}$. In that case

$$f^{(n)}\left(\frac{1}{2}\right) = \frac{n!}{\left(\frac{1}{2}\right)^{n+1}} = 2^{n+1} n!$$

This then leads to a Taylor expansion of $f(z)$ around $z = \frac{1}{2}$.

$$\begin{aligned}\frac{1}{1-z} &= 2 + 2^2\left(z - \frac{1}{2}\right) + 2^3\left(z - \frac{1}{2}\right)^2 + \cdots + 2^{n+1}\left(z - \frac{1}{2}\right)^n + \cdots \\ &= 2\left[1 + (2z-1) + \cdots + (2z-1)^n + \cdots\right]\end{aligned}$$

and as $f(z)$ is analytic within a circle of radius $\frac{1}{2}$ around $z_0 = \frac{1}{2}$, the Taylor series converges for $\left|z - \frac{1}{2}\right| < \frac{1}{2}$, i.e., for $|2z - 1| < 1$.

(c) The Taylor series in (b) can also obtianed directly using that in (a). Note

$$\begin{aligned}\frac{1}{1-z} &= \frac{1}{\frac{1}{2} + \left(\frac{1}{2} - z\right)} = \frac{2}{1 + (1-2z)} \equiv \frac{2}{1-w} \\ &= 2\left[1 + w + w^2 + \cdots + w^n + \cdots\right]\end{aligned}$$

and must converge for $|w| \equiv |2z - 1| < 1$.

7.6 Laurent Series

Theorem 7.5 Let a function $f(z)$ have an isolated singularity at $z = z_0$, enclosed in a circle $C_1 : |z - z_0| = R_1$ (see Fig. 7.7). Let C_2 be a circle with center z_0 of raidus $R_2 > R_1$ such that $f(z)$ is analytic on C_1, C_2 and within the annular region between C_1 asnd C_2. Then for any z in the annular domain, $f(z)$ is represented by the series

$$f(z) = \sum_{n=0}^{\infty} a_n (z - z_0)^n + \sum_{n=1}^{\infty} b_n (z - z_0)^{-n}. \tag{7.31}$$

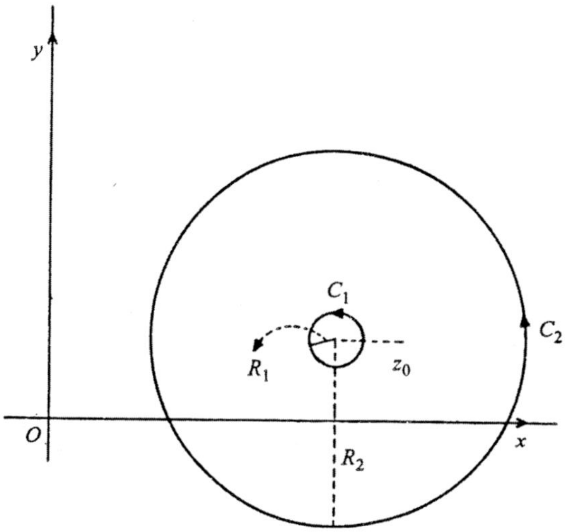

Fig. 7.7 *Annular domain for Laurent expansion*

Here a_n and b_n are given by the contour integrals

$$a_n = \frac{1}{2\pi i} \oint_{C_2} \frac{f(z)dz}{(z - z_0)^{n+1}} \qquad n = 0, 1, \ldots \tag{7.32a}$$

$$b_n = \frac{1}{2\pi i} \oint_{C_1} \frac{f(z)dz}{(z - z_0)^{-n+1}} \qquad n = 1, 2, \ldots \tag{7.32b}$$

Proof: We apply the Cauchy's integral formula for the multiply-connected domain enclosed between C_1 and C_2. We note that $f(z)$ is analytic on and within the boundary of this domain. The boundary of this domain is C_2 (taken anticlockwise) and $-C_1$ (i.e., C_1 in the reversed sense.) We then obtain, for any z with $R_1 < |z - z_0| < R_2$,

$$f(z) = \frac{1}{2\pi i}\oint_{C_2}\frac{f(\xi)d\xi}{\xi - z} - \frac{1}{2\pi i}\oint_{C_1}\frac{f(\xi)d\xi}{\xi - z}.$$

Both integrals are taken over contours in the anticlockwise sense.

Now, we shall deal with each of these two contour integrals in a manner similar to that in the proof of the Taylor's Theorem in Section 7.5. In the integrand of the first integral, we expand (noting $|\xi - z_0| > |z - z_0|$ on C_2),

$$\frac{1}{\xi - z} = \frac{1}{(\xi - z_0) - (z - z_0)} = \frac{1}{\xi - z_0}\left[1 + \frac{z - z_0}{\xi - z_0} + \left(\frac{z - z_0}{\xi - z_0}\right)^2 + \cdots\right.$$

$$= \frac{1}{\xi - z_0}\left[\sum_{n=0}^{N-1}\frac{(z - z_0)^n}{(\xi - z_0)^n} + \frac{(z - z_0)^N}{(\xi - z)(\xi - z_0)^{N-1}}\right].$$

When expanded, the first integral thus gives a series in nonnegative powers of $(z - z_0)$ up to $(N - 1)$th power and a residue $R_N(z)$.

In the integrand of the second integral over C_1, we expand, (noting $|\xi - z_0| < |z - z_0|$ on C_1)

$$\frac{1}{\xi - z} = \frac{-1}{(z - z_0) - (\xi - z_0)}$$

$$= -\frac{1}{(z - z_0)}\left[\sum_{n=0}^{N-1}\left(\frac{\xi - z_0}{z - z_0}\right)^n + \frac{(\xi - z_0)^N}{(z - z_0)^{N-1}(z - \xi)}\right].$$

The second integral when expanded gives a series in negative powers of $(z - z_0)$ up to $(z - z_0)^{-N}$ and a residue $R'_N(z)$. In the limit $N \to \infty$, $R_N(z)$ and $R_N(z')$ can be shown to vanish along the lines of the proof of Taylor's Theorem. Then the expansion of (7.31) with the coefficients given by (7.32) is established. The proof is left as an exercise (see Problem 7.20).

Examples

(i) Let $f(z) = \frac{1+2z}{z^2}$.

$f(z)$ has a singularity (only) at $z = 0$. Hence expansion of $f(z)$ around $z = 0$ will be a Laurent series (and not a Taylor series), i.e., containing negative powers of z as well. We can write

$$f(z) = \frac{2}{z} + \frac{1}{z^2} \qquad z \neq 0.$$

The above is already in the form of a Laurent series with all $a_n = 0$, $b_1 = 2$, $b_2 = 1$, all $b_n = 0$, $n \geq 3$. As we shall see the Laurent series of a function about an isolated singularity is *unique*; hence the above *must* be the Laurent series for $f(z)$ around $z = 0$. Alternately, one may also evaluate a_n and b_n more directly using

$$\oint_C z^{-n-1}dz = 2\pi i\,\delta_{n0} \quad (n \text{ integer})$$

and arrive at this result (C above is any contour enclosing $z = 0$ and which is in the domain of analyticity of $f(z)$). The above Laurent expansion holds in the entire complex plane except $z = 0$.

(ii) Let $f(z) = \frac{1}{z(z-1)}$; $\quad z_0 = 1$.

$f(z)$ has singularities at $z = 0$ and $z = 1$. We want a Laurent expansion of $f(z)$ about $z = 1$. As remarked in the example (i) above, from the result on the uniqueness of the Laurent expansion, *whatever series* we obtain in the integral powers of $(z - 1)$ will be the correct Laurent series. Hence we seek to expand $f(z)$ as

$$\begin{aligned} f(z) &= -\frac{1}{z} + \frac{1}{z-1} \\ &= \frac{-1}{1-(-z+1)} + \frac{1}{z-1} \\ &= -\sum_{n=0}^{\infty}(1-z)^n + \frac{1}{z-1} \qquad |1-z| < 1 \\ &= \frac{1}{z-1} + \sum_{n=0}^{\infty}(-)^{n-1}(z-1)^n. \end{aligned}$$

The above then gives the Laurent series for $f(z)$ around $z = 1$ and it converges in the region $0 < |z - 1| < 1$; which also is the largest annular domain with $z = 1$ as center in which $f(z)$ is analytic. (Note the singularity at $z = 0$.)

(iii) $f(z) = e^{1/z}$, $z = 0$ is a singularity. We can write down the Laurent series for $f(z)$ around $z = 0$ knowing the expansion of e^w.

$$f(z) = \sum_{n=0}^{\infty} \frac{1}{n!}\frac{1}{z^n}.$$

7.6.1 Alternate form for the coefficients a_n and b_n

Let C be any simple closed contour that lies in the annular region between C_1 and C_2 and which is taken in the anticlockwise sense. Then using the Cauchy–Goursat theorem we can re-express a_n and b_n alternatively as:

$$a_n = \frac{1}{2\pi i}\oint_{C_2}\frac{f(z)\,dz}{(z-z_0)^{n+1}} = \frac{1}{2\pi i}\oint_{C}\frac{f(z)\,dz}{(z-z_0)^{n+1}}, \tag{7.33}$$

$$b_n = \frac{1}{2\pi i}\oint_{C_1}\frac{f(z)\,dz}{(z-z_0)^{-n+1}} = \frac{1}{2\pi i}\oint_{C}\frac{f(z)\,dz}{(z-z_0)^{-n+1}}, \tag{7.34}$$

where use has been made of the fact that the integrand in (7.33) viz, $f(z)(z-z_0)^{-n-1}$ is analytic on C_2, C and within the multiply-connected region between C_2 and C; similarly, the integrand in (7.34), viz. $f(z)(z-z_0)^{n-1}$ is analytic on C, C_1 and the region in between.

We could then write

$$f(z) = \sum_{n=-\infty}^{\infty} c_n (z-z_0)^n \tag{7.35}$$

with

$$c_n = \frac{1}{2\pi i}\oint_C \frac{f(z)}{(z-z_0)^{n+1}} dz \tag{7.36}$$

where now all the coefficients are expressed in a unified manner as contour integrals over the *same* closed contour C.

7.6.2 Uniqueness of Laurent series

Let $f(z)$, analytic in the disc $0 < |z-z_0| < R$, be represented by a series

$$f(z) = \sum_{n=-\infty}^{\infty} C_n (z-z_0)^n \tag{7.37}$$

convergent everywhere in the domain $0 < |z-z_0| < R$. Then (7.33) is *the (unique) Laurent* series for $f(z)$ in this domain.

To see this we evaluate $\oint_C f(z)(z-z_0)^{-m-1}dz$, where C is a circular contour $|z-z_0| = R_1 < R$. From (7.33) we have,

$$\oint_C f(z)(z-z_0)^{-m-1} dz = \sum_{n=-\infty}^{\infty} C_n \oint_C (z-z_0)^n \ (z-z_0)^{-m-1} dz. \tag{7.38}$$

Here we have made an important assumption that the infinite sum over n on the R.H.S. and the contour integration can be interchanged. The justification for this will be mentioned later (see Section 7.7).

Now recall the result

$$\oint_C (z-z_0)^{-n-m-1} dz = \delta_{nm}\, 2\pi i. \tag{7.39}$$

We thus have

$$C_n = \frac{1}{2\pi i}\oint_c f(z)(z-z_0)^{-n-1} dz. \tag{7.40}$$

But these are precisely the coefficients in the Laurent series according to (7.36) of the previous subsection.

This proves the uniqueness of the series representation of $f(z)$ in powers of $(z - z_0)$ in the region of analyticity of $0 < |z - z_0| < R$; this series being always the Laurent expansion constructed earlier.

7.6.3 Nature of an isolated singularity

Consider a function $f(z)$ that has an isolated singularity at $z = z_0$. By definition, this means that $f(z)$ is analytic in some disc $|z - z_0| < \varepsilon_1$ except at $z = z_0$ $(\varepsilon_1 > 0)$. Now we enclose the singularity at $z = z_0$ in a smaller concentric circle $|z - z_0| = \varepsilon_2 < \varepsilon_1$. Then the conditions for the Laurent expansion are fulfilled in the annular region $\varepsilon_2 < |z - z_0| < \varepsilon_1$. Hence we can write the Laurent series that converges in this annular region:

$$f(z) = \sum_{n=0}^{\infty} a_n (z - z_0)^n + \sum_{n=1}^{\infty} b_n (z - z_0)^{-n}. \tag{7.41}$$

And as ε_2 can be taken to be arbitrarily small, (7.41) is valid for $0 < |z| < \varepsilon_1$. In absence of a singularity at $z = z_0$, the function $f(z)$ would have a Taylor expansion valid in $|z - z_0| < \varepsilon_1$ which would have only the nonnegative powers of $(z - z_0)$. The behaviour of $f(z)$ near the singularity $z = z_0$ is determined by the remaining terms involving b_n called the *principal part of* $f(z)$. There are three distinct classes of behaviour depending on three distinct possibilities regarding the set $\left\{b_n; n = 1, \dots\right\}$.

(a) All b_n's vanish.
(b) Only a finite number of b_n's are nonvanishing.
(c) An infinite number of b_n's are nonvanishing.

In the case (a), the limit of $f(z)$ as $z \to z_0$ exists and equals a_0. A redefinition then allows the function to be analytic in the entire disc $|z - z_0| < \varepsilon_1$; In other words, the newly defined function $\tilde{f}(z)$

$$\begin{aligned} \tilde{f}(z) &= f(z) \qquad && z \neq z_0 \\ &= a_0 && z = z_0 \end{aligned}$$

is analytic in the entire disc $|z - z_0| < \varepsilon_1$; and $z = z_0$ is a regular point of $\tilde{f}(z)$. In this case, we say that $f(z)$ has a removable singularity at $z = z_0$. (See Example (1) below).

In the case (b), it is clear that there will exist m such that $b_m \neq 0$ and $b_n = 0, \quad n > m$. Then $\phi(z) \equiv (z - z_0)^m \ f(z)$ has only a removable singularity at $z = z_0$. Thus the function

$$\begin{aligned} \tilde{\phi}(z) &= (z - z_0)^m f(z) \qquad && z \neq z_0 \\ &= b_m && z = z_0 \end{aligned}$$

is analytic in the entire disc $|z - z_0| < \varepsilon_1$. In this case $f(z)$ is said to have a pole of order m at $z = z_0$.

In the case (c), we cannot find m such that $b_n = 0,\ \ n > m$. In this case, the function $f(z)$ "diverges faster than any power" of $\left(\frac{1}{z-z_0}\right)$ near $z = z_0$. In such a case, $f(z)$ is said to have an essential singularity at $z = z_0$.

Definition: In the Laurent series (7.41), the coefficeint b_1 is called the *residue* of $f(z)$ at the isolated singular point $z = z_0$.

Example (3) Consider $f_1(z) = \frac{\sin z}{z}$. The function $f_1(z)$ is not defined at $z = 0$. The domain of definition of the function is the entire complex plane except the origin. The point $z = 0$ is a singular point of $f(z)$. However, we could define $\tilde{f}$ as:

$$\begin{aligned} \tilde{f}_1(z) &= \frac{\sin z}{z} \qquad & z \neq z_0, \\ &= 1 & z = z_0. \end{aligned}$$

Then $\tilde{f}(z)$ is analytic everywhere including the origin. Thus modification in the definiton of the function has removed the singularity at $z = 0$. The singularity at $z = 0$ is called a removable singularity.

Example (4) $f_2(z) = \frac{e^z}{z} = \frac{1}{z} + \sum\limits_{n=1}^{\infty} \frac{z^{n-1}}{n!}$.

The R.H.S. represents the (unique) Laurent expansion of $f(z)$ around $z = 0$. Here all $b_n^{\prime}s$ except b_1 vanish. Hence $f_2(z)$ has a pole of order 1 with residue 1.

Example (5) $f_3(z) = e^{1/z} = \sum\limits_{n=0}^{\infty} \frac{1}{n!z^n}$.

Here all $b_n^{\prime}s$ (infinite in number) in the above Laurent expansion are nonvanishing. Hence $z = 0$ is an essential singularity of $f_3(z)$.

7.7 Convergence

7.7.1 Uniform and absolute convergence of a complex power series

An analytic function may be represented by a Taylor series which may converge in a disc $|z - z_0| < R$ of radius R (not necessarily finite)

$$f(z) = \sum_{n=0}^{\infty} c_n (z - z_0)^n. \tag{7.42}$$

Such a series representation of a function would not be of too much use in practice if it was not possible to carry out operations such as differentiation and integration on the series term by term. Establishing these properties depends on the property of the above series (7.42) called "uniform convergence". We shall now give definitions of uniform and absolute convergence.

Definitions:

1. A series $\sum_{n=0}^{\infty} a_n z^n$ is said to converge at z to a limit f when, for any ε, however small, one can find N_0 such that for any $N > N_0$,

$$\left| f - \sum_{n=0}^{N} a_n z^n \right| < \varepsilon \quad \text{for} \quad N > N_0. \tag{7.43}$$

   Of course, f in the above equation may depend on z.
   If, for the above series at a given z, f cannot be found with the above property, the series is said to diverge at z.
2. The set of points z where the series converges constitutes the domain of convergence D.
3. In the domain of convergence D, the series $\sum a_n z^n$ converges to the function $f(z)$.
4. When one considers equation (7.43) over the entire domain of convergence D (and not at just a single point), one may generally expect that for a given ε, the number N_0 may differ from point to point. If, however, a single N_0 can be found for all $z \in D_1 \subset D$ for which (7.43) holds, we say that $\sum_{n=0}^{\infty} a_n z^n$ converges to $f(z)$ *uniformly* in domain D_1.
5. A series $\sum_{n=0}^{\infty} a_n z^n$ is said to be *absolutely convergent* at z iff the series $\sum_{n=0}^{\infty} |a_n z^n|$ converges there.

Example To illustrate the above definitions, we consider the familiar series

$$S(z) = \sum_{n=0}^{\infty} z^n. \tag{7.44}$$

As we have already seen, the above series is the Taylor expansion of $\frac{1}{1-z}$ about $z = 0$ and converges for $|z| < 1$. Thus, $S(z)$ converges to $f(z) = \frac{1}{1-z}$ for $|z| < 1$.

The *interior* of the unit circle is the domain of convergence D.

The series does *not* converge at *any* point on the unit circle (see Problem 7.14). The series $\sum_{n=0}^{\infty} |z|^n$ also converges $\left[\text{to } \frac{1}{1-|z|}\right]$ within D. Hence $\sum z^n$ is "absolutely convergent" within D.

Further, let z_1 be any point within the unit circle (however close to the circle). Let $|z_1| = \lambda < 1$. Then, since

$$\sum_{n=0}^{N} z^n = \frac{1 - z^{N+1}}{1 - z} = \frac{1}{1 - z} - \frac{z^{N+1}}{1 - z} \qquad (z \neq 1),$$

we have,

$$\left| \frac{1}{1 - z} - \sum_{n=0}^{N} z^n \right| = \frac{|z|^{N+1}}{|1 - z|} \leq \frac{|z|^{N+1}}{|1 - |z||}.$$

So, for *any* z with $|z| < \lambda < 1$,

$$\left| f(z) - \sum_{n=0}^{N} z^n \right| < \frac{\lambda^N}{|1 - \lambda|}.$$

Thus, given ε however small, we can choose $N_0 \equiv \frac{\ln[\varepsilon(1-\lambda)]}{\ln \lambda}$ *independent of* z, such that

$$\left| f(z) - \sum_{n=0}^{N} z^n \right| < \varepsilon \qquad \text{for } N > N_0;\ |z| < \lambda.$$

Hence, according to the definition of uniform convergence, $S(z)$ converges *uniformly* within the circle $|z| < \lambda < 1$.

Now we establish the following result.

Theorem 7.6 Let the series $\sum_{n=0}^{\infty} a_n z^n$ be convergent at $z = z_0$. Then

(i) it is absolutely convergent for all z such that $|z| < |z_0| \equiv R$,
(ii) it is convergent for all z such that $|z| < |z_0|$,
(iii) it is uniformly convergent for all z such that $|z| < |z_1| < R$ (where $|z_1|$ can however be arbitrarily close to R).

Proof:

(i) The series $\sum_{n=0}^{\infty} a_n z_0^n$ is convergent. Hence there exists a positive number M such that $|a_n z_0^n| < M$ for all n (why?). Now consider the series $\sum a_n z^n$. We shall first show that the above series is absolutely convergent for $|z| < R$. Let $|z| = \lambda R$; $0 \leq \lambda < 1$. Now,

$$\sum_n |a_n z^n| = \sum_n |a_n z_0^n| \left| \frac{z}{z_0} \right|^n = \sum_n |a_n z_0^n| \lambda^n < \sum_n M \lambda^n = M \frac{1}{1 - \lambda} < \infty.$$

Thus the series $\sum_n |a_n z^n|$ of nonnegative numbers is bounded from above and hence must converge. (It cannot oscillate as terms are positive and hence add up.) This proves (i).

(ii) Further, writing $a_n z^n = Z_n$, we note that the two series $\sum$ Re Z_n and $\sum$ Im Z_n must converge by comparison test. This is so since Re $Z_n \leq |Z_n|$ and Im $Z_n \leq |Z_n|$ and $\sum |Z_n|$ is known to converge (from part (i)). Hence the series $\sum_{n=0}^{\infty} Z_n$ converges, since for the partial sum

$$\sum_{n=0}^{N} Z_n = \sum_{n=0}^{N}(\text{Re } Z_n + i \text{ Im } Z_n) = \left[\sum_{n=0}^{N} \text{Re } Z_n\right] + i\left[\sum_{n=0}^{N} \text{Im } Z_n\right]$$

and each of the series on the R.H.S. converges to a limit. (For a more subtle treatment see Problem 7.33.) This prove (ii).

(iii) Now let $\sum_{n=0}^{\infty} a_n z^n$ converge to $f(z)$ for $|z| < R$. We want to establish that the convergence is uniform in the disc $|z| < |z_1| < R$ where we have chosen some z_1 with $|z_1| < R$. For this we want to show that given ε however small, one can find $N_0(\varepsilon)$ *independent of* z such that for every z with $|z| < |z_1|$

$$\left|f(z) - \sum_{n=0}^{N-1} a_n z^n\right| = \left|\sum_{n=N}^{\infty} a_n z^n\right| < \varepsilon \qquad \text{for any } N > N_0(\varepsilon).$$

To see this, consider

$$\sum_{n=N}^{\infty} a_n z^n = \lim_{m\to\infty} \sum_{n=N}^{m} a_n z^n$$

and

$$\left|\sum_{n=N}^{m} a_n z^n\right| \leq \sum_{n=N}^{m} |a_n z^n| \leq \sum_{n=N}^{m} |a_n z_1^n|.$$

Now the series $\sum_{n=0}^{\infty} |a_n z_1^n|$ is convergent by (i). Hence given ε, however small, there is $N_0(\varepsilon)$ possibly depending only on z_1 such that $\sum_{n=N}^{\infty} |a_n z_1^n| < \varepsilon$ whenever $N > N_0(\varepsilon)$.

Hence,

$$\lim_{m\to\infty}\left|\sum_{n=N}^{m} a_n z^n\right| \leq \lim_{m\to\infty}\sum_{n=N}^{m} |a_n z_1^n| < \varepsilon$$

whenever $N > N_0(\varepsilon)$. And this $N_0(\varepsilon)$ is independent of z for any z such that $|z| < |z_1|$. (Further $|z_1|$ can be taken to be arbitrarily close to $|z_0| = R$.) This proves (iii).

Comments: We now comment on the contents of the theorem. Let $\tilde{z}$ be the point with the minimum modulus (it it exists) such that the series $\sum_{n=0}^{\infty} a_n \tilde{z}^n$ fails to converge. Then the series cannot converge for any z with $|z| > |\tilde{z}| = R$. For if it were to converge at z with $|z| = R_1 > R$; then, according to the part (ii) of the theorem, it must converge everywhere within the circle $|z| < R_1$, including the point $\tilde{z}$, leading to a contradiction. Thus the series $\sum_{n=0}^{\infty} a_n z^n$ converges everywhere within the circle $|z| < R$ (by assumption) and nowhere outside $|z| > R$. The circle $|z| = R$ is called the *circle of convergence* of the series $\sum_{n=0}^{\infty} a_n z^n$ and R is called its *radius of convergence*. If no such point $\tilde{z}$ exists then the radius of convergence is infinite.

Further, according to the part (i) of the theorem, the series is absolutely convergent within the circle of convergence. And according to the part (iii) of the theorem it converges uniformly in any concentric circle interior to the radius of convergence.

The behaviour of the series on the circle of convergence differs from series to series. There must however be at least one point on the circle of convergence where the series $\sum a_n z^n$ does not converge. (For example see Problem 7.14.)

7.7.2 Interchange of summation and differentiation/integration

Now we state two important consequences of uniform convergence of the series $\sum_{n=0}^{\infty} a_n z^n$ within the circle of convergence $|z| < R$. These state that the series can be differentiated term by term or can be integrated term by term. Thus, let $\sum_{n=0}^{\infty} a_n z_n$ converge to $f(z)$ for $|z| < R$. Then $\sum_{n=0}^{\infty} a_n n z^{n-1}$ converges, within $|z| < R$, to $f'(z)$. Further, let C be a (finite) contour within the disc $|z| < R$ and $g(z)$ be a continuous function. Then $\int_C g(z)dz \sum a_n z^n$ can be evaluated by term by term addition of the integrals involved. In other words, the sum $\sum_{n=0}^{\infty} a_n \int_C dz g(z) z^n$ converges to $\int_C g(z) f(z) dz$.

Theorem: Let $\sum_{n=0}^{\infty} a_n z^n$ converge to $f(z)$ for $|z| < R$. Then $f'(z)$ exists and the series $\sum_{n=0}^{\infty} a_n n z^{n-1}$ converges to $f'(z)$ for all z with $|z| < R$.

Proof: We can write for $|z| < R$,

$$f(z) = \sum_{n=0}^{N} a_n z^n + R_N(z). \tag{7.45}$$

From uniform convergence of the series, it follows that given ε however small, one can choose $N_0(\varepsilon)$ independent of z such that for all z with $|z| < R$

$$|R_N(z)| < \varepsilon \qquad \text{for } N > N_0(\varepsilon). \tag{7.46}$$

Now consider the quantity

$$\frac{R_N(z+\Delta z) - R_N(z)}{\Delta z} \tag{7.47}$$

Now, both $|R_N(z+\Delta z)| < \varepsilon$ and $|R_N(z)| < \varepsilon$ (ε begin independent of z). Thus when $|z| < R$ and $|z + \Delta z| < R$ and $N > N_0$ on account of the uniform convergence,

$$\left|\frac{R_N(z+\Delta z) - R_N(z)}{\Delta z}\right| \le \frac{|R_N(z+\Delta z)| + |R_N(z)|}{|\Delta z|} = \frac{2\varepsilon}{|\Delta z|}. \tag{7.48}$$

Now we shall show that the L.H.S. of (7.48) can be arbitrarily small by choosing N sufficiently large and Δz sufficiently small. In (7.48) we know that we can choose N_0 sufficiently large, so that for any given Δz, $\varepsilon(N_0) = C|\Delta z|^2$, C being a fixed constant. Thus, for any given Δz and $N > N_0(|\Delta z|)$,

$$\left|\frac{R_N(z+\Delta z) - R_N(z)}{\Delta z}\right| < 2C|\Delta z|.$$

Now for any Δz,

$$\left|\frac{f(z+\Delta z) - f(z)}{\Delta z} - \frac{\sum_{n=0}^{N} a_n(z+\Delta z)^n - \sum a_n z^n}{\Delta z}\right|$$
$$= \left|\frac{R_N(z+\Delta z) - R_N(z)}{\Delta z}\right| < 2C|\Delta z| \qquad N > N_0(|\Delta z|),$$

and can be made arbitrarily small by taking Δz sufficiently small (provided N is chosen as mentioned.) Taking the limit $\Delta z \to 0$ in the above equation,

$$\lim_{\Delta z \to 0} \frac{f(z+\Delta z) - f(z)}{\Delta z} = \sum_{n=0}^{\infty} a_n\, n\, z^{n-1}$$

so that this limit exists and

$$\frac{d}{dz} \sum_{n=0}^{\infty} f_n(z) \equiv \frac{d}{dz} \sum a_n z^n = \sum_{n=0}^{\infty} \frac{d}{dz} f_n(z),$$

thus establishing that term by term differentation is valid.

Generalizations:

(i) As the series $\sum\limits_{n=0}^{\infty} a_n n z^{n-1} \equiv \sum b_n\, z^n$ converges for $|z| < R$, repeated applications of the theorem leads to the existence of $f^{(n)}(z)$ and expressions for it.

(ii) A minor modification of the above proof leads to following general result: Given a sequence of analytic functions $f_n(z)$, let $\sum\limits_{n=0}^{\infty} f_n(z)$ converge uniformly to $f(z)$ in a domain $|z| \in D$. Then in D, $f(z)$ is analytic and $\sum\limits_{n=0}^{\infty} f_n'(z)$ converges to $f'(z)$.

Now a similar result will be given for the term-by-term integration of an infinite series.

Theorem: Let the series $\sum\limits_{n=0}^{\infty} a_n z^n$ converge within the disc $|z| < R$. Let $g(z)$ be continuous for $|z| < R$. Then for,any (finite) contour C within the disc,

$$\int_C g(z) \sum_{n=0}^{\infty} a_n z^n dz = \sum_{n=0}^{\infty} a_n \int_C g(z) z^n dz.$$

Proof: Consider

$$\left| \int_C g(z) \sum_{n=0}^{\infty} a_n z^n dz - \int_C g(z) \sum_{n=0}^{N} a_n z^n dz \right| \le \left| g(z) \left\{ \sum_{n=0}^{\infty} a_n z^n - \sum_{n=0}^{N} a_n z^n \right\} \right|_{\max} \cdot L$$

using Darboux theorem (L is the length of the contour C). Now as $g(z)$ is continuous in the disc $|z| < R$, there exists M such that

$$|g(z)| < M \text{ for } |z| < R.$$

Further, as the given series converges,

$$\left| \sum_{n=0}^{\infty} a_n z^n - \sum_{n=0}^{N} a_n z^n \right| < \varepsilon(N),$$

where $\varepsilon(N)$ can be made arbitrarily small by taking N arbitrarily large. Thus,

$$\left| g(z) \left\{ \sum_{n=0}^{\infty} a_n z^n - \sum_{n=0}^{N} a_n z^n \right\} \right|_{\max} \cdot L < M\varepsilon(N)L$$

$$\longrightarrow 0 \qquad \text{as } N \longrightarrow \infty$$

Thus,

$$\left| \int_C g(z) \sum_{n=0}^{\infty} a_n z^n dz - \sum_{n=0}^{N} a_n \int_C g(z) z^n dz \right| \to 0 \text{ as } N \to \infty.$$

This implies

$$\int_C g(z) \sum_{n=0}^{\infty} a_n z^n \, dz = \sum_{n=0}^{\infty} a_n \int_C g(z) z^n dz,$$

i.e., the term-by-term integration can be carried out in the L.H.S.

7.8 Residue Theorem

7.8.1 Residues at poles

I) When one attempts to find the residue of a function $f(z)$ at a pole (of mth order) at $z = z_0$, the function can in some cases be written *explicitly* in the form

$$f(z) = \frac{\phi(z)}{(z - z_0)^m} \qquad z \neq z_0, \tag{7.49}$$

where $\phi(z)$ is analytic in some neighbourhood of $z = z_0$ and $\phi(z_0) \neq 0$. $\phi(z)$ can be Taylor-expanded around $z = z_0$ in the disc $|z - z_0| < \varepsilon$ where it is analytic. We then can write a Laurent series for $f(z)$:

$$f(z) = \frac{1}{(z - z_0)^m} \left\{ \phi(z_0) + \phi'(z_0)(z - z_0) + \cdots + \frac{\phi^{(n)}(z_0)}{n!}(z - z_0)^n + \cdots \right\}$$

$$= \frac{\phi(z_0)}{(z - z_0)^m} + \frac{\phi'(z_0)}{(z - z_0)^{m-1}} + \cdots + \frac{\phi^{(m-1)}(z_0)}{(m-1)!(z - z_0)} + \cdots$$

The coefficient of $\frac{1}{z - z_0}$ in the above series is, by definition, the residue

$$R(z_0) = \frac{\phi^{(m-1)}(z_0)}{(m-1)!} \tag{7.50}$$

This result is very useful in practice.

Examples

(i) Let $f(z) = \frac{z+2}{(z+5)(z-1)^3}$.
$f(z)$ has poles at $z = 1$ (third order) and $z = -5$ (first order). To find the residue of $f(z)$ at $z = -5$, the simple pole, we note

$$f(z) = \frac{\phi_1(z)}{z+5} \qquad \text{with} \qquad \phi_1(z) = \frac{(z+2)}{(z-1)^3}.$$

We note that $\phi_1(z)$ is analytic in a disc $|z+5| < 6$ (as the singularity of $\phi_1(z)$ nearest to $z = -5$ is at $z = 1$). We also note that $\phi_1(z_0) = \phi_1(-5) \neq 0$.
Then the residue of $f(z)$ at $z = z_0 = -5$ is

$$R(-5) = \phi_1(-5) = \frac{3}{6^3} = \frac{1}{72}.$$

On the other hand, to find the residue of $f(z)$ at the pole of the third order at $z = 1$, we note that $f(z)$ can be written as

$$f(z) = \frac{\phi_2(z)}{(z-1)^3} \qquad \text{with} \qquad \phi_2(z) = \frac{z+2}{z+5}$$

and $\phi_2(z_0) \neq 0$. $\phi_2(z)$ is analytic in the disc $|z-1| < 6$. The residue of $f(z)$ at $z = 1$ is then

$$R(1) = \frac{\phi_2''(1)}{2!} = -\frac{1}{72}.$$

II) There are situtations where one has to find the residue at a pole $z = z_0$ of a function f that does not *explicitly* fall in the form of $f(z)$ in (7.49). Consider for example the case when $f(z)$ has the form of a quotient of analytic functions

$$f(z) = \frac{P(z)}{Q(z)}, \tag{7.51}$$

where $P(z)$ and $Q(z)$ are both analytic at $z = z_0$ and $Q(z)$ has a zero of order m at $z = z_0$. (If $Q(z)$ can *explicitly* be written as $(z - z_0)^m G(z)$ with $G(z)$ analytic at z_0, then (7.51) becomes a special case of (7.49) with $\phi(z) = P(z)/G(z)$). In other cases, the result of (7.50) cannot be directly applied. For example, consider the function

$$f(z) = \frac{\cos z}{\sin^2 z} \equiv \frac{P(z)}{Q(z)}.$$

Clearly, $P(z)$ and $Q(z)$ are both entire and $Q(z)$ has a zero of order two at $z = 0$ (in particular).
Now,

$$Q(z) = \sin^2 z = z^2 \frac{\sin^2 z}{z^2}$$

is not yet in the form $z^2G(z)$ where $G(z)$ is analytic at $z = 0$. But

$$G(z) = \frac{\sin^2 z}{z^2};\ z \neq 0$$

has a removable singularity at $z = 0$; and the function $G(z) = \dfrac{\sin^2 z}{z^2}$ $(z \neq 0)$ and $G(0) = 1$ is analytic everywhere.

Thus, in general, if $Q(z)$ has a zero at $z = z_0$ of order m, it can be written, after removing the removable singularity at $z = z_0$ as (see Problem 7.34)

$$Q(z) = (z - z_0)^m\ G(z), \tag{7.52}$$

where $G(z)$ is analytic at $z = z_0$ and $G(z_0) \neq 0$.

Then we have

$$f(z) = \frac{P(z)}{(z - z_0)^m G(z)} \equiv \frac{\phi(z)}{(z - z_0)^m} \tag{7.53}$$

and the results of the previous subsection directly apply formally provided $P(z_0) \neq 0$. These results, however, are not easy to use directly. We therefore prefer to express the results in terms of the analytic functions P and Q and their derivatives.

In the case of a small m, much simpler and direct results can be given. For example, for the case $m = 1$,

$$R(z_0) = \phi(z_0) = \frac{P(z_0)}{G(z_0)}$$

but as is easily seen

$$Q'(z_0) = \frac{d}{dz}Q(z)\bigg|_{z=z_0} = \frac{d}{dz}[(z - z_0)\ G(z)]\bigg|_{z=z_0} = G(z_0).$$

Thus,

$$R(z_0) = \frac{P(z_0)}{Q'(z_0)}. \tag{7.54}$$

In a similar manner, one can show that for $m = 2$,

$$\begin{aligned} R(z_0) &= \phi'(z_0) \\ &= \frac{P'(z_0)}{G(z_0)} - \frac{P(z_0)}{G^2(z_0)}\,G'(z_0) \\ &= \frac{2P'(z_0)}{Q''(z_0)} - \frac{2P(z_0)}{3[Q'''(z_0)]^2}\,Q'''(z_0), \end{aligned} \tag{7.55}$$

a result entirely in terms of P and Q and its derivatives.

Solved example 7.1 Find the residues of $f(z) = \tan z$ as its poles.

$$f(z) = \frac{\sin z}{\cos z} \equiv \frac{P(z)}{Q(z)}$$

has poles at $z = \left(n + \frac{1}{2}\right)\pi$. These are simple poles as these are the locations of simple zeros of $\cos z$. Note that $\frac{d}{dz}(\cos z) = -\sin z$ is nonvanishing at these points.

Hence, for $z_0 = \left(n + \frac{1}{2}\right)\pi$,

$$R\left[\left(n + \frac{1}{2}\right)\pi\right] = \frac{P(z_0)}{Q'(z_0)} = \frac{\sin z_0}{-\sin z_0} = -1$$

for all integral n.

Solved example 7.2 Find the residue of $f(z) = \dfrac{\cos z}{\sin^2 z}$ at $z = 0$.

Here $P(z) = \cos z$, $Q(z) = \sin^2 z$. $Q(z)$ has a zero of order two at $z = 0$. Hence $R(0)$ is obtained from (7.55). Noting $P'(0) = 0$, $Q'''(0) = 0$, we find

$$R(0) = 0.$$

This result could have been read off noting that $f(z)$ is an even function of z. (Please see Problem 7.26.)

III) A third and more direct method for finding the residue is to write down, wherever easy enough, a Laurent series for $f(z)$ around its pole at $z = z_0$. This can be conveniently done when $f(z)$ is a product or quotient of a number of factors such that the Taylor series, each of which can easily be written down. Only the relevant term in the Laurent series, viz, the $(z - z_0)^{-1}$ term, is all that is needed and its coefficient gives $R(z_0)$. The method is best illustrated by a few examples.

Solved example 7.3 Find the residue of $f(z) = (\coth^2 z)(1 + z)$ at its pole.

$$f(z) = \frac{\cosh^2 z(1 + z)}{\sinh^2 z}.$$

Here, $f(z)$ has (only) a pole of order two coming from the zero of $\sinh^2 z$ at $z = 0$. We know the Taylor expansions of $\cosh z$ and $\sinh z$ around $z = 0$ and these can be used to write a Laurent series for $f(z)$ around $z = 0$. Ultimately, we need here only the coefficient of $\frac{1}{z}$ in this series.

$$f(z) = \left[1 + \frac{z^2}{2!} + \frac{z^4}{4!} + \cdots\right]^2 (1 + z)\left[z + \frac{z^3}{3!} + \cdots\right]^{-2}$$

$$= z^{-2}\left[1 + \left(\frac{z^2}{2!} + \frac{z^4}{4!} + \cdots\right)\right]^2 (1 + z)\left[1 + \left(\frac{z^2}{3!} + \cdots\right)\right]^{-2}$$

We now need to expand the two square brackets in binomial expansions. In view of the z^{-2} in the front, we need only the coefficient of z coming from the product of the three factors. The result then is easily seen to be

$$f(z) = z^{-2}\left[1 + z + O(z^2)\right] = \frac{1}{z^2} + \frac{1}{z} + \text{nonnegative powers of } z.$$

The above series for $f(z)$ must be the Laurent series for $f(z)$ around $z = 0$ as the latter is unique and thus

$$R(0) = 1.$$

Solved example 7.4 Locate the poles, their order and find the residue at these poles for

$$f(z) = \frac{1}{z\ \sin z\pi\left(z + \frac{1}{2}\right)}.$$

This problem illustrates the efficacy of all the three methods in three different situations.

$f(z)$ has poles at $z = 0$, $z = -\frac{1}{2}$, $z = n (n \neq 0)$ where n is an integer.

As $\sin z\pi$ has simple zeros at $z = n(n = 0, \pm 1, \ldots)$ [note: $\cos n\pi \neq 0$], $f(z)$ has simple pole at $z = \pm 1, \pm 2, \ldots,$; and a double pole at $z = 0$.

It also has a simple pole at $z = -\frac{1}{2}$.

(i) In the neighbourhood of $z = -\frac{1}{2}$, we can write

$$f(z) = \frac{\phi(z)}{z + \frac{1}{2}} \quad \text{with} \quad \phi(z) = \frac{1}{z \sin \pi z}$$

and hence

$$R\left(-\frac{1}{2}\right) = \phi\left(-\frac{1}{2}\right) = 2.$$

(ii) In the neighbourhood of $z = n$ (n is a nonzero integer)

$$f(z) = \frac{p(z)}{q(z)}$$

where $p(z) \equiv \frac{1}{z\left(z+\frac{1}{2}\right)}$ and $q(z) = \sin \pi z$ are analytic at $z = n(n \neq 0)$. $q(z)$ has a simple zero at $z = n$ and hence the residue at $z = n(n \neq 0)$ is

$$R(n) = \frac{p(n)}{q'(n)} = \frac{\frac{1}{n\left(n+\frac{1}{2}\right)}}{\pi \cos n\pi} = \frac{(-)^n}{\pi n\left(n + \frac{1}{2}\right)}; \qquad n \neq 0.$$

(iii) Now remains the double pole at $z = 0$. Note now that $f(z)$ cannot be written explicitly as

$$f(z) = \frac{\phi(z)}{z^2},$$

where $\phi(z)$ is (strictly) analytic at $z = 0$. Nor can it be written in the form

$$f(z) = \frac{p(z)}{q(z)}$$

with

$$q(z) = z^2 r(z).$$

For $r(z) = \frac{\sin \pi z}{z}$ is singular at $z = 0$, and thus not analytic at $z = 0$. However, $r(z) = \frac{\sin \pi z}{z}$ has only a removable singularity at $z = 0$. In fact, we could alternately use the series

$$\begin{aligned} z \sin \pi z &= z \left\{ \pi z - \frac{(\pi z)^3}{3!} + \frac{(nz)^5}{5!} - \cdots \right\} \\ &= \pi z^2 \left\{ 1 - \frac{(\pi z)^2}{3!} + \frac{(nz)^4}{5!} - \cdots \right\} \\ &\equiv z^2 \tilde{r}(z). \end{aligned}$$

$\tilde{r}(z)$ now representing the series is known to converge (why?) for all z. Thus $\tilde{r}(z) = r(z)$, $z \neq 0$ and $\tilde{r}(0) = \pi$. Thus $\tilde{r}$ is obtained form r by removing singularity at $z = 0$. Now, we have

$$f(z) = \frac{p(z)}{z^2 \tilde{r}(z)} \equiv \frac{\tilde{\phi}(z)}{z^2} \qquad \text{with } \tilde{\phi}(z) = \frac{1}{\left(z + \frac{1}{2}\right) \tilde{r}(z)}$$

with $\tilde{\phi}(z)$ analytic near $z = 0$. Then,

$$R(0) = \tilde{\varphi}'(0) = -\frac{4}{\pi}.$$

(iv) A yet better way to obtain $R(0)$ is to write down a few terms in the Laurent series of $f(z)$ about $z = 0$.

Note:

$$\frac{1}{z}\frac{1}{z+\frac{1}{2}}\frac{1}{\sin z\pi} = \frac{1}{z}\frac{2}{1+2z}\frac{1}{\pi z - \frac{(\pi z)^3}{3!}+\dots}$$

$$= \frac{1}{z}\cdot 2\cdot\left[1-2z+(2z)^2-\dots\right]\frac{1}{\pi z}\left[1-\frac{\pi^2 z^2}{3!}+\frac{\pi^4 z^4}{5!}-\right]^{-1}$$

$$= \frac{2}{\pi z^2}\left[1-2z+4z^2-\dots\right]\left[1+\left(\frac{\pi^2 z^2}{3!}-\frac{\pi^4 z^4}{5!}+\dots\right)+O(z^4)\right]$$

$$= \frac{2}{\pi z^2}-\frac{4}{\pi z}+ \text{nonnegative powers of } z.$$

This immediately yields

$$R(0) = -\frac{4}{\pi}$$

with much greater ease.

7.8.2 Residue theorem

Consider a function $f(z)$ analytic in a domain D except for an isolated singularity at $z = z_0 \in D$. It can then be expanded in a Laurent series valid in disc $|z-z_0| < \varepsilon$ within D with $\varepsilon > 0$. (The disc in fact extends to the nearest point of the boundary of D).

$$f(z) = \sum_{n=0}^{\infty} a_n(z-z_0)^n + \sum_{n=0}^{\infty} b_n(z-z_0)^{-n-1}.$$

Now consider the integral $\oint_C f(z)dz$, over a circle $C : |z-z_0| = R < \epsilon$, taken anticlockwise.

Using the result regarding the interchange of summation and integration (see Section 7.7.2), we can write

$$\oint_C f(z)dz = \sum_{n=0}^{\infty} a_n \oint_C (z-z_0)^n dz + \sum_{n=0}^{\infty} b_n \oint_C (z-z_0)^{-n-1} dz.$$

Now, as has been established earlier, all the contour integrals above except one vanish:

$$\oint_C (z-z_0)^{-n-1} dz = \delta_{n,0} 2\pi i.$$

Thus,

$$\oint_C f(z)dz = 2\pi i b_1. \tag{7.56}$$

Applying the Cauchy's theorem, we can see that the above result, in fact, holds for any closed contour C within D that encloses this singularity. Thus, the result for $\oint_C f(z)dz$ for any such C depends only on the residue of $f(z)$ at z_0, i.e., the coefficient of $\frac{1}{z-z_0}$ term in the Laurent series.

This result can now be straightforwardly generalized to the case when the closed contour C (within the domain D) contains any (finite) number of isolated singularities.

Theorem: Let $f(z)$ be analytic on and within a closed simple contour C (taken anti-clockwise) except for a finite set of isolated singularities $z_1, z_2, \ldots, z_n$ with residues $b_1, b_2, \ldots, b_n$. Then,

$$\oint_C f(z)dz = 2\pi i \sum_{i=1}^{n} b_i.$$

Proof: Since each of z_i is an isolated singularity, we can enclose it within a circle C_i : $|z - z_i| = \varepsilon_i$ so that C_i encloses only one singularity z_i of $f(z)$. We take the circles C_i in the clockwise sense. Now from the result (7.56) established earlier (and taking due account of change in the sense of the contours),

$$\oint_{C_i} f(z)dz = -2\pi i b_i. \tag{7.57}$$

Now we can apply the Cauchy–Gourset theorem to the multiply-connected region bounded by $C_1, \ldots, C_n$ and C. Then,

$$\oint_C f(z)dz + \sum_{i=1}^{n} \oint_{C_i} f(z)dz = 0. \tag{7.58}$$

Using (7.57) in (7.58) above, we have

$$\oint_C f(z)dz = 2\pi i \sum_{i=1}^{n} b_i.$$

7.9 Evaluations of Integrals using the Residue Theorem

The technique of evaluation of contour integrals using the residue theorem can be applied to evaluation of a number of types of real integrals and this often provides a superior method for their evaluation. The main ingredients of the procedure are :

(i) In the case of improper integrals, regard the integral over the real variable (taken as x) as a contour integral over (an interval of) the real axis of the complex plane taken as an open contour C_1; and in the case of integrals over a trigonometric funtion, regard the variable θ as going over a unit circle or its arc.

(ii) Try to obtain a closed contour C of which this interval over the real variable is a part. (Of course, not any closed contour would do, more on this later in (v).)

(iii) Generalize the integrand, if necessary, so that a contour integral over a closed contour C is defined over the complex plane, the segment of which over the real axis or arc of a unit circle gives the real integral to be evaluated.

(iv) Evaluate the closed contour integral using the Residue Theorem. This of course works provided the complex integrand has certain analyticity properties required of in the statement of the Residue Theorem.

(v) Evaluate the integral over remaining part C_2 of the contour ($C = C_1 + C_2$). Almost invariably, this procedure is useful where the contribution from the remaining part C_2 of the contour can be shown to vanish for the integrand in question or when it can be related to the original integral itself.

Let us now illustrate the above steps with the help of a number of examples.
Consider the real integrals

$$I_1 = \int_{-\infty}^{\infty} \frac{dx}{x^2+1} = 2\int_{0}^{\infty} \frac{dx}{x^2+1}; \tag{7.59}$$

$$I_2 = \int_{-\infty}^{\infty} \frac{\cos x}{x^2+1} dx = 2\int_{0}^{\infty} \frac{dx \cos x}{x^2+1}; \tag{7.60}$$

$$I_3 = \int_{0}^{2\pi} \frac{d\theta}{a+\cos\theta} \qquad a > 1. \tag{7.61}$$

I_1 and I_2 can be regarded as integrals over the entire real axis of the complex plane. This is C_1. I_3 can be regarded as an integral over a unit circle with $z = e^{i\theta}$, $d\theta = -ie^{-i\theta}dz$. For I_3, C_1 is the unit circle. This elaborates step (i).

For I_1 and I_2, a closed contour (shown in Fig. 7.8) with $R \to \infty$ in the end constitutes C. C_2 here is the semicircle at infinity. For I_3, C_1 is already a closed contour and C_1 itself is C. This illustrates step (ii).

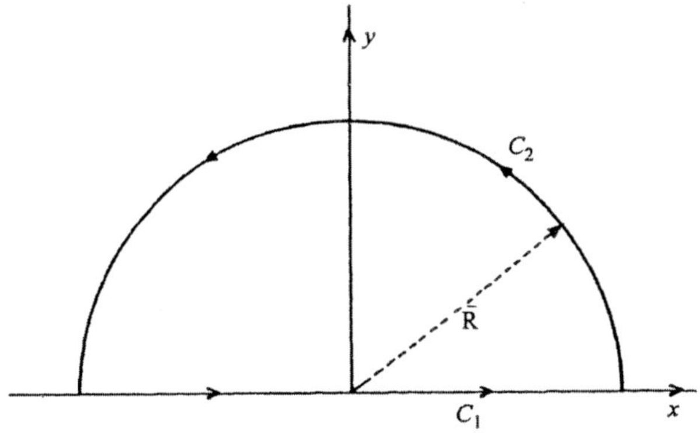

Fig. 7.8 *Closed contour for evaluation of I_1 and I_2*

Now we come to step (iii), where we generalize the integrand, if necessary. In the case of I_1, we generalize the integrand

$$\frac{1}{x^2+1} \longrightarrow \frac{1}{z^2+1}. \tag{7.62}$$

Then $\oint_C \frac{1}{z^2+1} dz$ contains I_1 as a contribution.

Here we note an important point. The generalization given by Eq. (7.62) above is not the only generalization of the integrand that yields I_1 as a part of the contour integral. Had we kept the integrand same,

$$\frac{1}{x^2+1} \longrightarrow \frac{1}{(\text{Re } z)^2+1} = \frac{1}{x^2+1}$$

or had we replaced

$$\frac{1}{x^2+1} \longrightarrow \frac{1}{[f(x,y)]^2+1} \qquad f(x,0) \equiv x,$$

we could have obtained I_1 as a part of a contour integral. However these generalizations do not possess the analytic properties that (7.62) enjoys. The analyticity of $\frac{1}{z^2+1}$ (except at $z = \pm i$) helps one evaluate the contour integral by the Residue Theorem which otherwise would have been difficult.

In the case of I_2, we could generalize

$$\frac{\cos x}{x^2+1} \longrightarrow \frac{\cos z}{z^2+1}.$$

Substituting for $\cos z$ via $2 \cos z = e^{iz} + e^{-iz}$, it turns out that the treatment of the two terms in the interval so obtained by this substitution differs for the contour at infinity. (More on this after Jordan Lemma). We, instead, note

$$\frac{\cos x}{x^2+1} = \text{Re}\left[\frac{e^{ix}}{x^2+1}\right]$$

and perform the replacement

$$\frac{e^{ix}}{x^2+1} \longrightarrow \frac{e^{iz}}{z^2+1}$$

which is analytic everywhere except at $z = \pm i$.

In the case of I_3, no replacement is needed; we only need to express $\cos\theta$ in terms of z:

$$\cos\theta = \frac{e^{i\theta}+e^{-i\theta}}{2} = \frac{z+1/z}{2} \qquad |z| = 1$$

in the integrand. We also note the replacement for integral measure and limits

$$\int_0^{2\pi} d\theta \ldots = -i \oint_C \frac{dz}{z} \ldots$$

The step (iv) then is a straightforward process using the Residue Theorem, and details differ from example to example. This step, in fact, completes the evaluation for I_3 since C and C_1 are in fact identical in this case.

In step (v), we must evaluate the integral over C_2 (for the first two examples). This step is in fact crucial for obtaining I_1 and I_2 from the respective contour integrals. One can show that in either case, integrals over C_2 vanish as $R \to \infty$. In the case of I_2, proof of this needs a special result which we shall prove later, viz. the Jordan Lemma.

We now put together the above steps and work out the integrals one by one.

(i) $$I_1 = \int_{-\infty}^{\infty} \frac{dx}{x^2+1} = \lim_{R\to\infty} \int_{-R}^{R} \frac{dx}{x^2+1}$$

$$= \lim_{R\to\infty} \left[\oint_C \frac{dz}{z^2+1} - \int_{C_2} \frac{dz}{z^2+1} \right].$$

Here C and C_2 are defined in Fig. 7.8. Now the integrand in the first contour integral has simple poles at $z = +i$, of which the one at $z = i$ is included in C for any $R > 1$. Thus,

$$\oint_C \frac{dz}{z^2+1} = 2\pi i R(i) = 2\pi i \frac{1}{2i} = \pi.$$

Hence,

$$I_1 = \pi - \lim_{R\to\infty} \int_{C_2} \frac{dz}{z^2+1}.$$

Now we note that as $|z^2+1| > |z^2| - 1 = R^2 - 1$ on C_2, assuming $R > 1$,

$$\left| \int \frac{dz}{z^2+1} \right| \le \frac{1}{R^2-1} L = \frac{\pi R}{R^2-1}$$

$$\longrightarrow 0 \qquad \text{as } R \longrightarrow \infty.$$

Thus, $I_1 = \pi$.

Here we remark that in this example I_1 happens to be a simple integral for which the result is already known. But much more complicated integrals can also be done by this method as easily as I_1 by an identical procedure. In such cases this method proves to be mathematically superior.

(ii) $$I_2 = \int_{-\infty}^{\infty} \frac{\cos x}{x^2+1} dx = \text{Re} \int_{-\infty}^{\infty} dx \frac{e^{ix}}{x^2+1} = \text{Re} \lim_{R\to\infty} \int_{-R}^{R} \frac{e^{ix}}{x^2+1} dx$$

Now,

$$\int_{-R}^{R} dx \frac{e^{ix}}{x^2+1} = \oint_C \frac{e^{iz}}{z^2+1} dz - \int_{C_2} \frac{e^{iz}}{z^2+1} dz$$
$$= 2\pi i R(i) - \int_{C_2} \frac{e^{iz}}{z^2+1} dz.$$

Here the residue is found to be $R(i) = \frac{e^{-1}}{2i}$. Further, using Jordan Lemma (to be proved subsequently), we can show that (see next subsection 7.9.1)

$$\lim_{R\to\infty} \int_{C_2} \frac{e^{iz}}{z^2+1} dz = 0.$$

Thus,

$$I_2 = \int_{-\infty}^{\infty} \frac{\cos x}{x^2+1} dx = \text{Re} \int_{-\infty}^{\infty} \frac{e^{ix}}{x^2+1} = \text{Re} \left(\frac{\pi}{e}\right) = \pi/e.$$

(iii) $$I_3 = \int_0^{2\pi} \frac{d\theta}{a+\cos\theta} \qquad a > 1$$
$$= \int_0^{2\pi} \frac{d\theta}{a+\frac{1}{2}(e^{i\theta}+e^{-i\theta})}$$
$$= \oint_C \frac{dz/iz}{a+\frac{1}{2}(z+\frac{1}{z})} \qquad |z| = 1$$
$$= -2i \oint_C \frac{dz}{z^2+2az+1}.$$

Here C is the unit circle taken counterclockwise.
Now the integrand has poles at

$$z = -a \pm \sqrt{a^2-1}.$$

Of the two poles the one at $r_1 \equiv \left(\sqrt{a^2-1}-a\right)$ lies inside the unit circle while the other one is outside the unit circle. (The product of these two roots is one; hence the smaller one must then lie inside the unit circle). Hence,

$$I_3 = (-2i)(2\pi i)R(r_1) = \frac{2\pi}{\sqrt{a^2-1}}.$$

7.9.1 Jordan lemma

As we have stated earlier, the evaluation of certain integrals involving $e^{i\alpha z}$ over a semicircle of radius R (as $R \to \infty$) involves the use of the Jordan Lemma. In this subsection, we shall state and prove it.

Theorem: Let C_R denote a semicircle in the upper half of the complex plane (UHP) with center at the origin. Let $f(z)$ be a function that tends uniformly to zero with respect to $\theta = \arg z$ as $|z| \to \infty$ for $0 \leq \theta \leq \pi$. Then for α real and positive,

$$\lim_{R\to\infty} I_R \equiv \lim_{R\to\infty} \int_{C_R} e^{i\alpha z} f(z) dz = 0. \tag{7.63}$$

Comments: Before we give the formal proof, we shall make comments. Consider first the simpler integral with $f(z)$ replaced by 1. Then we have the integral

$$I_R^0 = \int_{C_R} e^{i\alpha z} dz \qquad \alpha > 0. \tag{7.64}$$

Now, note that

$$e^{i\alpha z} = e^{i\alpha(R\cos\theta + i\sin\theta)} \equiv e^{-\alpha R\sin\theta} e^{i\alpha R\cos\theta} \tag{7.65}$$

Thus as $R \to \infty$, the integrand is exponentially damped for $0 < \theta < \pi$ as $\sin\theta > 0$ in this range. The length of the contour increases as πR, i.e. only linearly as R. One may then naively expect the integral to vanish exponentially rapidly as $R \to \infty$. This, however, proves incorrect, the reason being that as θ approaches 0 or π, the exponential damping $e^{-\alpha R\sin\theta}$ is turned off. For a given R, the angular width of this region where exponential damping is absent is $\Delta\theta \sim \frac{1}{R\alpha}$ at both ends of the contour and here the integrand is $O(1)$. The length of the contour of this region at both ends is $2R\Delta\theta(R) \sim \frac{2}{\alpha}$, which is a constant even as $R \to \infty$. Thus, in this region $\Delta\theta(R)$ at both ends of the integration range, ends up giving a contribution of $O(1)$ to the integral. It is here that the condition that $|f(z)| \to 0$ as $R \to \infty$ uniformly becomes relevant to make I_R vanish as $R \to \infty$; without it the integral would not vanish. Thus, to summarize, I_R vanishes because

(i) contribution over "most" of the semicircle is exponentially damped. [Here, we must have $\alpha > 0$ *and* $0 \leq \theta \leq \pi$ (semicircle in the upper half plane), or $\alpha < 0$ *and* $\pi \leq \theta \leq 2\pi$ (semicircle in the lower half plane).]

(ii) the contribution from the "end regions" of the contour vanishes on account of the fact that $|f(z)| \to 0$ as $R \to \infty$ uniformly.

The role played by the condition, $|f(z)| \to 0$ *uniformly* as $R \to \infty$, is best seen by studying Problem 7.32.

Now we come to the mathematical proof.

Proof: The fact that $f(z)$ tends to zero uniformly implies that, *independent* of $\theta\,(0 \le \theta \le \pi)$, we can write on the contour

$$|f(z)| < \varepsilon(R) \qquad \text{where } \varepsilon(R) \to 0 \text{ as } R \to \infty. \tag{7.66}$$

Now, we note

$$I_R = \int_0^{\pi} d\theta (iR) e^{i\theta} f(Re^{i\theta}) e^{-\alpha R \sin\theta} e^{i\alpha R \cos\theta}$$

so that

$$\begin{aligned}
|I_R| &\le \text{integral of the modulus} \\
&= \int_0^{\pi} d\theta R |f(Re^{i\theta})| e^{-\alpha R \sin\theta} \\
&\quad < \varepsilon(R) \int_0^{\pi} d\theta R e^{-\alpha R \sin\theta} \qquad \text{using (7.66)} \\
&= 2R\varepsilon(R) \int_0^{\pi/2} d\theta e^{-\alpha R \sin\theta}
\end{aligned}$$

To put an upper bound on this integral we note that

$$\sin\theta \ge \frac{2\theta}{\pi} \qquad \text{for } 0 \le \theta \le \pi/2$$

so that

$$e^{-\alpha R \sin\theta} \le e^{-2\alpha R\theta/\pi} \qquad 0 \le \theta \le \pi/2.$$

Thus,

$$\begin{aligned}
|I_R| &\le 2R\varepsilon(R) \int_0^{\pi/2} d\theta e^{-2\frac{\alpha R}{\pi}\theta} \\
&= \varepsilon(R) \frac{\pi}{\alpha} (1 - e^{-\alpha R}) \\
&\longrightarrow 0 \quad \text{as} \quad R \longrightarrow \infty \qquad \text{using (7.66).}
\end{aligned}$$

This completes the proof.

Example Here we consider the integral in one of the examples discussed already in the previous section, viz.

$$\int_{C_R} \frac{e^{iz}}{z^2+1} dz.$$

We note that the conditions of the theorem are fulfilled here. Here $\alpha = 1 > 0$. Also for $R > 1$,

$$|f(z)| = \frac{1}{|z^2+1|} \leq \frac{1}{|z^2|-1} = \frac{1}{R^2-1} \quad \text{on } C_R.$$

Thus,

$$|f(z)| \leq \frac{1}{R^2-1} \equiv \varepsilon(R),$$

and ε is independent of θ and goes to zero as $R \to \infty$. Thus $|f(z)| \to 0$ uniformly as $|z| \to \infty$.

Thus, as advertised earlier,

$$\lim_{R\to\infty} \int_{C_R} \frac{e^{iz}}{z^2+1} dz = 0.$$

Modification of Jordan Lemma A trivial modification of the Jordan Lemma can be made in which the result holds also if (i) $\alpha < 0$ (ii) C_R is replaced by C'_R a semicircle in the LHP. Note, in this case also, $e^{i\alpha z} \to 0$ exponentially as $R \to \infty$ for $\pi < \theta < 2\pi$. The remaining considerations are identical.

Example It now straightforwardly follows that

$$\lim_{R\to\infty} \int_{C'_R} \frac{e^{-iz}}{z^2+1} dz \to 0.$$

7.9.2 Further real integrals using Residue Theorem

The classes of integrals discussed in the previous subsection cover broadly a large set of real integrals that can be evaluated using the residue theorem. There are, of course, several other integrals that can be evaluated by this method which require a special choice of contour for each case and thus cannot be classified easily. We will mention some of these at the end as examples and/or problems. There is, however, left another class of problems that can be given a general treatment and these involve complex integrands that have branch cuts. We now illustrate this class by an example.

Consider the real integral, defined for $0 < a < 1$,

$$I_4 = \int_0^\infty \frac{x^{-a}}{x+1} dx.$$

This can be looked upon as a contour integral

$$\int_{C_1} \frac{z^{-a}}{z+1} dz,$$

where (i) C_1 is the straight line contour along the positive real axis, (ii) z^{-a} has to be defined as in Subsection 6.4.4, providing its branch cut.

We now note several problems that arise. Firstly z^{-a} is not analytic at $z = 0$; hence C_1 must start from $z = \varepsilon$; ε real and positive. Now one must find a closed contour C, of which C_1 is a part, to which residue theorem can be applied.

We note that C should have the following properties.

(a) It must not cross the branch cut.
(b) It must not contain a part in the *finite* complex plane where the integral over which it is unrelated to I_4 (or difficult to evaluate).
(c) The contour must avoid $z = 0$, as the integrand is singular there.

The only choice available turns out to be

- As the branch cut can be chosen arbitrarily, choose it along the positive real axis.
- Choose the contour to pass just below the branch cut, circling the origin and then just above the branch cut. Finally close it at infinity. This is shown in Fig. 7.9.

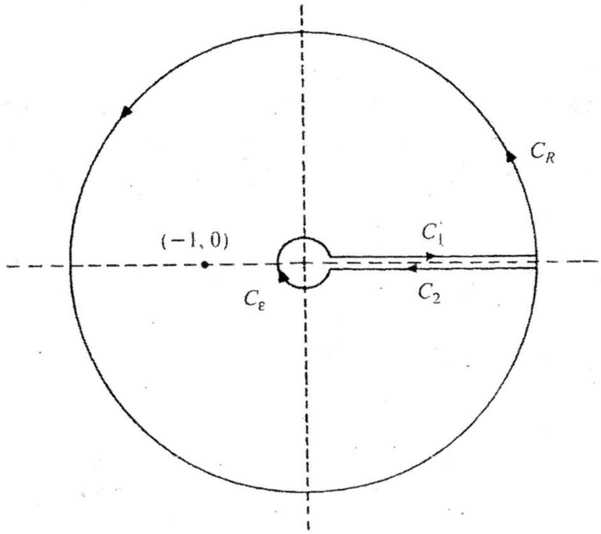

Fig. 7.9 *Contour for the evaluation of I_4*

We note how our objective is now accomplished.

(i) Integrand is analytic on and within C (except for pole at $z = -1$).
(ii) Integration over C_1 is I_4.
(iii) There are two more segments on C in the finite complex plane: C_2 and C_ε (see figure). The integrand over C_2 differs from the integrand over C_1 because these are on two different sides of the branch cut of z^{-a}. *But* they are proportional.

In fact,

$$\left.\frac{z^{-a}}{z+1}\right|_{C_2} = \left[e^{-2\pi i a}\right] \left.\frac{z^{-a}}{z+1}\right|_{C_1}$$

at neighbouring points. Hence the integral over C_2 is proportional to I_4. This is crucial to the choice of C.

(iv) Further, integral over C_ε can be shown to vanish as $\in \to 0$. The same holds for integral over C_R as $R \to \infty$.

(v) Now, application of the residue theorem gives integral over C and this (as $\in \to 0$ and $R \to \infty$) is then directly proportional to I_4.

Now, we give the complete derivation. We define

$$z^{-a} = e^{-a \log z}$$

where we choose the branch cut for $\ln z$ along the positive real axis, so that

$$\log z = \text{Log}\, r + i\theta \qquad 0 \leq \theta < 2\pi.$$

We shall also now specify the contours C_1 and C_2 more carefully. For mathematical reasons, it is convenient to regard C_1 as making a small positive angle δ with the x-axis and then consider the limit $\delta \to 0^+$ in the end. Similarly, we regard C_2 as making a small negative angle $-\delta$ with the x-axis (and finally consider the limit $\delta \to 0$). Thus, on C_1 and C_2

$$\begin{aligned} z^{-a} &= r^{-a} e^{-ai\delta} && \text{on } C_1 \quad \varepsilon < r < \infty, \\ z^{-a} &= r^{-a} e^{-ai(2\pi-\delta)} && \text{on } C_2 \quad \varepsilon < r < \infty. \end{aligned}$$

(These expressions are much simpler mathematically than what we would have if C_1 and C_2 were straight lines parallel to the x-axis at a small distance Δ.)

Now the integrand is analytic on and within C except at the simple pole at $z = -1$. Hence by the residue theorem, for $R > 1$,

$$\oint_C \frac{z^{-a}}{z+1} dz = 2\pi i R(-1) = 2\pi i e^{-\pi i a}.$$

Also,

$$\int_{C_1} \frac{z^{-a}}{z+1} dz = e^{i\delta(1-a)} \int_\varepsilon^R \frac{r^{-a} dr}{re^{i\delta}+1} \to \int_\varepsilon^R \frac{r^{-a} dr}{r+1} \quad \text{as} \quad \delta \to 0;$$

$$\int_{C_2} \frac{z^{-a}}{z+1} dz = e^{i(1-a)(2\pi-\delta)} \int_R^{\varepsilon} \frac{r^{-a} dr}{re^{-i\delta}+1}$$

$$\rightarrow -e^{-2\pi i a} \int_{\varepsilon}^{R} \frac{r^{-a} dr}{r+1} \qquad \text{as } \delta \rightarrow 0;$$

$$\left| \int_{C_\epsilon} \frac{z^{-a}}{z+1} dz \right| = \left| \int_{\delta}^{2\pi-\delta} \frac{\varepsilon^{-a+1} e^{-ia\theta} e^{i\theta} i d\theta}{\varepsilon e^{i\theta}+1} \right|$$

$$\leq \frac{\varepsilon^{1-a}}{1-\varepsilon} 2\pi \qquad 0 < \varepsilon < 1;$$

where we have used $|1 + \varepsilon e^{i\theta}| > 1 - \varepsilon$ for $0 < \varepsilon < 1$.

Thus,

$$\int_{C_\varepsilon} \frac{z^{-a}}{z+1} dz \rightarrow 0 \qquad \text{as } \varepsilon \rightarrow 0 \qquad \text{as } 1 > a.$$

Further, noting that on C_R,

$$\left| \frac{z^{-a}}{z+1} \right| \leq \frac{R^{-a}}{R-1} \qquad \text{for } R > 1,$$

$$\left| \int_{C_R} \frac{z^{-a}}{z+1} dz \right| < \frac{R^{-a}}{R-1} 2\pi R = \frac{2\pi R^{1-a}}{R-1}$$

$$\rightarrow 0 \qquad \text{as } R \rightarrow \infty \qquad \text{since } a > 0.$$

Now,

$$\oint_C = \oint_{C_1} + \oint_{C_2} + \oint_{C_\varepsilon} + \oint_{C_R} = 2\pi i e^{-\pi i a}.$$

Now we take the limits $\delta \rightarrow 0$, $\varepsilon \rightarrow 0$ and $R \rightarrow \infty$ successively on both sides. We, thus, obtain

$$\int_0^{\infty} \frac{r^{-a} dr}{r+1} - e^{-2\pi i a} \int_0^{\infty} \int_0^{\infty} \frac{r^{-a} dr}{r+1} = 2\pi i e^{-\pi i a}$$

$$\left[1 - e^{-2\pi i a}\right] I_4 = 2\pi i e^{-\pi i a}.$$

This leads to $I_4 = \frac{\pi}{\sin \pi a}$.

Problems

On § 7.1

7.1 Let

$$\left.\begin{array}{l} x = -\cos t \\ y = -\cos 2t \end{array}\right\} 0 \le t \le \pi$$

$$\left.\begin{array}{l} x = 2 + \cos t \\ y = -2 + \cos 2t \end{array}\right\} \pi \le t \le 2\pi.$$

Sketch the curve in the x–y plane. Is the curve (i) an arc, (ii) a simple arc, (iii) a regular arc, (iv) a contour, (v) a closed contour?

7.2 Draw the curve

$$\left.\begin{array}{l} x = \sin 2t \\ y = \sin t \end{array}\right\} 0 \le t \le 2\pi.$$

Is this a simple arc? Smooth arc?

7.3 Prove the result of (7.12) § 7.3.

7.4 Prove that

$$\oint_C z^{-n-1} dz = 2\pi i \delta_{n0}$$

where C is any contour enclosing the origin in anticlockwise sense.

7.5 Consider the integral $I = \int_C e^z z^2 dz$ where C represents a contour joining points (0, 0) to $(1, 0)$ and $(1, 0)$ to (1,1). Using Cauchy's theorem and Darboux inequality, show that $|I| < 2e\sqrt{2}$.

On § 7.2

7.6 Verify by direct evaluation that $\oint_C z^2 dz$ is identically zero, where C represents (i) the unit circle, (ii) the square with corners at $(1, 1)$, $(1, -1)$, $(-1, -1)$ and $(-1, +1)$. Comment in the light of Cauchy–Goursat theorem.

7.7 (a) Consider a small rectangular contour of sides ε_1 and ε_2 parallel to the axes. Evaluate approximately $\oint f(z)dz$ across the contour up to $O(\varepsilon_1\varepsilon_2)$. Show how vanishing of this integral is directly linked to the CR conditions and certain assumptions of continuity.

(b) Assuming the result in (a) for an analytic function f, show $\oint_C f(z)dz = 0$ over any simple closed contour C. Do so by dividing the area enclosed in terms of rectangular cells.

[Hint: Suppose the area is broken up in squares, the number of squares $\simeq$ area/ε^2 contribution to integral along one square $\sim O(\varepsilon^3)$, etc.]

7.8 Prove the Greens theorem for a contour C that is a square of side ε parallel to the axes up to $O(\varepsilon^2)$. Hence prove it for an arbitrary closed simple contour C.

7.9 Evaluate the integral $\oint_C dz \frac{e^z}{z}$ over a square contour C (in anticlockwise sense) whose corners are $(1, 1), (-1, +1), (-1, -1), (1, -1)$. What would the result be if the side of the square is halved in size, keeping it centered at a origin? Repeat the above exercise for $\oint_C dz e^z z^{-n}$ where n is an integer greater than one.

On § 7.3

7.10 Find the value of

$$\oint_C \frac{e^z \sin z}{z^3} dz$$

where C represents a unit circle in the positive sense.

7.11 Verify the Cauchy integral-formula for $f(z) = z^2, z_0 = 1$ and C is the unit circle $|z-1| = 1$. [Hint; Express $z = 1 + e^{i\phi}$ with $-\pi < \phi \leq \pi$].

7.12 Prove the result (7.18) of §7.4 for $f^{(n)}(z)$.

7.13 Let $f(z)$ be a function defined in a simply-connected domain D. Consider a function

$$F(z; C) = \int_{z_0}^{z} f(z')\, dz'$$

where the integral on the R.H.S. is taken over an open contour C from z_0 to z lying within D. [Note that generally F will depend on C.]

(a) Show that $F(z, C)$ is independent of C for any such contour C iff

$$\oint_{C'} f(z')\, dz' \equiv 0 \tag{7.1}$$

for every closed contour C' within D.

(b) Assuming that the condition (i) is fulfilled, show that

$$\frac{dF}{dz} = f(z)$$

for z within D; and hence that $F(z)$ is analytic in D.

(c) Argue further that $f(z)$ must be analytic in D.(This is Morera's theorem).

(d) Consider the function $f(z) = \frac{1}{z}$ and let D be the entire complex plane leaving out −ve real axis and the origin. Apply the above construction to obtain $F(z)$ and verify that it is analytic in D. Why is it necessary to leave out an entire line through origin in defining D and not just the singularity at $z = 0$?

On § 7.4

7.14 (a) Show that the Taylor series

$$\frac{1}{1-z} = 1 + z + z^2 + \cdots + z^n + \cdots$$

does not converge for any point on the unit circle (which is also the circle of convergence).

(b) Show, however, that the series

$$\ln(1+z) = \sum_{n=1}^{\infty} \frac{(-)^{n-1} z^n}{n}$$

converges at all points on its circle of convergence (where is it?) except at the obvious singularity at $z = -1$.

7.15 Obtain the Taylor series for $\ln z$ around $z = +1$. What is its radius of convergence?

7.16 Obtain the Taylor series for $\frac{z+1}{z(z-2)}$ around $z = 1$ and $z = \frac{1}{2}$. Obtain the radius of convergence in each case and relate it to the singularities of the function.

7.17 Taylor expansion of $f(z) = \frac{1}{\sin \pi z}$ is to be made about a point $z = x_0$ on the real axis. What are the largest and the smallest radii of convergence as x_0 is varied?

7.18 Let $f(z) = u(x, y) + iv(x, y)$ be analytic in a domain D.

(a) Show then (using the fact that $f^{(n)}(z)$ exist in D for all n) all partial derivatives of u and v exist in D.

(b) Write down the Taylor series for the real functions u and v (of two variables) around $z = z_0 \in D$. Keep terms upto order $(x-x_0)^2$, $(y-y_0)^2$, $(x-x_0)(y-y_0)$.

(c) Using CR conditions, and the above expansions show how the expansion for $f(z)$ so obtained agrees with that from Taylor's theorem upto an appropriate order.

(d) Apply the above procedure to $f(z) = \frac{1}{1-z}$ and $z_0 = 0$.

7.19 Show that $f^{(n)}(z_0) = \dfrac{\partial^n u}{\partial x^n}(x_0, y_0) + i\dfrac{\partial^n v}{\partial x^n}(x_0, y_0)$ where $f(z)$ is an analytic function at $z = z_0$. Give alternate expressions for $f^{(n)}(z_0)$. Construct an argument to show that all the $2(n+1)$ partial derivatives of the nth degree of u and v can be related to the only two independent derivatives $\dfrac{\partial^n u}{\partial x^n}$ and $\dfrac{\partial^n v}{\partial x^n}$.

On § 7.6

7.20 Show that $R_N(z)$ and $R'_N(z)$ defined in the proof for Laurent theorem vanish as $N \to \infty$.

7.21 Determine the singular points (if any) and their nature for

(i) $f_1(z) = z\mathrm{cosec} z$,

(ii) $f_2(z) = z \cosh\left(\frac{1}{z}\right)$,

(iii) $f_3(z) = x = \mathrm{Re}\, z$,

(iv) $f_4(z) = \dfrac{\sin z}{(z+1)^4}$.

7.22 Find Laurent series for

$$f(z) = \mathrm{cosec}\, z \qquad \text{about } z = 0$$
$$= \mathrm{cosec}\, z \qquad \text{about } z = \pi$$

and hence determine the residue at respective singular points.

7.23 Find a few terms of the Laurentz series for

$$f(z) = \frac{\mathrm{cosec}\, z}{z+1} \text{ about } z = 1.$$

On § 7.8

7.24 Find the residues at the poles of

$$f(z) = \frac{z^2+2}{z^2(z^2+1)}.$$

7.25 Locate the poles and their order and find the residues at these poles for

$$f(z) = \frac{1}{z(\cos \pi z)\left(z+\frac{1}{2}\right)^2}.$$

7.26 Find by inspection the residue of

$$f(z) = \frac{\cos z}{\sin^2 z}$$

at $z = 0$. [Hint : Note that $f(z)$ is even in z. What then can you say about the coefficeint of $\frac{1}{z}$ in its Laurent series about $z = 0$?]. Hence, deduce the residue of $f(z)$ at the poles $z = 2n\pi$. [Hint : Note that $f(z)$ is periodic in z with period 2π]. How would you obtain the residue at $z = (2n+1)\pi$ [Hint: $f(z+\pi) = -f(z)$].

7.27 (a) Deduce the result of (7.55) of § 7.8 by completing the algebra.

(b) Deduce the result of (7.55) of § 7.8 as follows:

(i) Express P and Q as Taylor series around $z = z_0$ keeping a few terms. Note that Q has zero of order two at z_0.

(ii) Evaluate the first two terms in the Laurent series for the quotient $P(z)/Q(z)$, and pick out the coefficient of $1/(z - z_0)$.

7.28 Consider a set of n line charges, each parallel to the z-axis, and carrying uniform linear charge densities $\lambda_1, \ldots, \lambda_n$ respectively. Let C be a closed curve in the x–y plane.

(a) Show using Gauss' law that the electric field $\mathbf{E}$ of these charges obeys

$$\oint_C \mathbf{E} \cdot \hat{n}\, dl = \frac{1}{2\varepsilon_0} \sum_{i \in C} \lambda_i,$$

where the sum goes over line charges enclosed by C.

(b) Show that the electric field of a line charge at (x_0, y_0) viz. $\mathbf{E} = (E_x, E_y, 0)$ can be associated with a function $f_0(z) = \frac{\lambda_0}{(z - z_0)2\pi\varepsilon_0}$ such that $E_x = \text{Re}\, f(z)$ and $E_y = -\text{Im}\, f(z)$.

(c) Show that for the general case in (a) we can associate $\mathbf{E}$ with a function f:

$$\mathbf{E} \to f(z) = \sum_{i=1}^{n} \frac{\lambda_i}{(z - z_i)2\pi\varepsilon_0}.$$

(d) Show then that the Gauss's law in (a) is then identical with

$$\text{Im} \oint_C f(z) dz = 2\pi$$

(sum over residues of $f(z)$ enclosed).

(e) Show that $\nabla \cdot \mathbf{E} = 0$ corresponds to the CR condition $u_x = v_y$ for $f(z)[\equiv u + iv]$ associated with $\mathbf{E}$; and $\nabla \times \mathbf{E} = 0$ corresponds to the other CR condition $u_y = -v_x$.

(f) The magnetic field due to an infinite straight current carrying wire (current I) is given by

$$\mathbf{B} = \frac{\hat{k} \times \mathbf{r}}{r^2} \frac{\mu_0 I}{2\pi}.$$

Show that you can associate a function $f(z) = \frac{\mu_0 I}{2\pi} \frac{1}{z}$ with this $\mathbf{B}$ such that $\text{Re}\, f(z) = B_x$, $\text{Im}\, f(z) = -B_y$. Hence show that the Ampere's law

$$\oint_C \mathbf{B} \cdot \mathbf{dl} = \mu_0 I_{\text{encl}}$$

can be cast in the form of the residue theorem

$$\mathrm{Re}\left[\oint_C f(z)\,dz - 2\pi i\,(\text{Residue at the pole within } C)\right] = 0.$$

7.29 Evaluate

(i) $$I_n = \int_0^{2\pi} \frac{\cos n\theta d\theta}{5 - 4\cos 2\theta}$$

(ii) $$I = \int_0^{2\pi} \frac{d\theta}{a + b\,\sin\theta} \qquad |b| < |a|$$

(iii) $$I = \int_0^{2\pi} \frac{d\theta}{a + b\sin^2\theta},\ I' = \int_0^{2\pi} \frac{d\theta}{a + b\cos^2\theta}$$

(iv) $$I = \int_0^{2\pi} \cos^{2m+2}\theta\,\sin^2\theta d\theta$$

On § 7.9

7.30 Evaluate the integral $\int_0^{\infty} e^{ix^2} dx$ by the following procedure:

(i) Define the integral as $\lim_{\varepsilon\to 0} \int_0^{\infty} e^{(i-\varepsilon)x^2} dx$

(ii) Show that the integral can be written as $A \int_{C_1} e^{z^2} dz$, where A is a constant and C_1 goes over the straight line $z = Re^{\pi i/4} e^{i\varepsilon/2}$; $(\varepsilon \to 0)$.

(iii) By considering the closed contour that contains C_1, C_0 (positive real axis) and the $\frac{1}{8}$-circle at infinity, evaluate the result.

(iv) Hence establish the general result

$$\int_{-\infty}^{\infty} e^{iax^2+2ibx} dx = \sqrt{\frac{\pi i}{a}} e^{-ib^2/a}$$

7.31 Determine the range of 'a' for which the real integral

$$I = \int_0^{\infty} \frac{x^{-a}\tanh x}{x^2+1} dx; \qquad a \text{ real}$$

is defined. Evaluate I by an application of the residue theorem.
[Hint: Choose a set of enlarging discrete square contours with $|\tanh z|$ bounded on each.]

7.32 Study the role played by the condition that $|f(z)| \to 0$ *uniformly*, in the statement of the Jordon Lemma, by considering

$$f(z) = e^{-2izz^*/z-z^*} \qquad y \neq 0,$$
$$= 0 \qquad y = 0 \;\; x \neq 0.$$

Show that for $|z| = R \to \infty$, $f(z) \to 0$ as $R \to \infty$ in UHP, but it does not approach zero uniformly. Show further that

$$\int_0^{1/\alpha R} e^{-\alpha R \sin\theta} |f(z)| \, d\theta R$$

does not in fact go to zero as $R \to \infty$. Do so by establishing a lower bound on the integral. Hence argue that

$$\tilde{I}_R = \int_0^{1/\alpha R} e^{i\alpha z} f(z) \; e^{i\theta} \; iR \; d\theta$$

also does not vanish as $R \to \infty$. Hence argue further that I_R itself does not vanish as $R \to \infty$.
[Hint: (i) $\frac{2\theta}{\pi} \leq \sin\theta \leq \theta$ for $0 \leq \theta \leq \pi/2$;
(ii) establish that $I_R - 2\tilde{I}_R$ vanishes exponentially as $R \to \infty$.

Miscellaneous

7.33 Show explicitly that if for a sequence $(Z_1, Z_2, \ldots, Z_n, \ldots)$, $\mathrm{Re}Z_n$ and $\mathrm{Im}Z_n$ approach limits X and Y, then

$$\lim_{n\to\infty} Z_n = X + iY.$$

7.34 Show *directly* that if a series $\sum_{n=0}^{\infty} a_n z^n$ converges for $z = z_1$, then it must converge for all points on the ray from 0 to z_1, i.e. for all z such that $z = \alpha z_1, \; 0 \leq \alpha < 1$.

8 Complex Variables – III

This chapter is devoted to a number of special isolated topics in the theory of functions of complex variables that are of interest to a physicist. These are

(a) Analytic continuation
(b) Conformal mapping and applications to electrostatics
(c) Method of steepest descent
(d) Multivalued functions and Riemann surfaces

8.1 Analytic Continuation

The number of strong results derived in the last chapter on analytic functions point to the fact that the behaviour of an analytic function in different parts of its domain of definition is strongly correlated. For example, the Cauchy–Goursat theorem states that for an analytic function $f(z)$, the contour integral over *any* contour with domain of analyticity D is zero:

$$\oint_C f(z)dz = 0.$$

Cauchy integral formula allows the determination of the function at every point interior to a closed contour C, given its values on the contour C, if the function is known to be analytic. We have also seen that an analytic function is infinitely differentiable and its derivatives of all orders are themselves analytic. These results suggest that the behaviour of an analytic function over the domain of its definition must be highly restricted in order to enable such strong results to hold. Hence, we may expect that knowing the analytic function over a small region in the domain of its definition should enable us to say something about it in another region (possibly remote) in the (connected) domain of its definition. In fact, a very strong result turns out to be true: the former can, in fact *uniquely determine* the latter in a simply-connected domain. This is the basis of analytic continuation.

For comparison, we may note that no such results are possible for functions of a real variable. For example, if one considers a differentiable function $f(x)$ of a single variable x (i.e., $f'(x)$ exists) over an interval (a, b), it does not even guarantee the existence of and the values of second and higher derivatives over (a, b). In particular, given the values of a function $f(x)$ in $a < x < b$ and given that it is differentiable in (a, b), no conclusion could be drawn about its values outside this interval (but in the domain of its definition). (See Problem 8.7.) One can go even further. Consider functions of a real variable that are infinitely differentiable. For example, consider the functions,

$$f(x) = 1 + x + x^2 + \ldots \qquad -1 < x < 1$$

which is infinitely differentiable in $-1 < x < 1$. One can then ask if one can construct a *unique* function that is infinitely differentiable except at singularities, and coincides with $f(x)$ defined about in $-1 < x < 1$? The answer is no. To see this consider

$$f_1(x) = \frac{1}{1-x} \qquad x \neq 1$$

and

$$f_2(x) = \frac{1}{1-x} + \theta(x-1)\frac{C}{x} \qquad x \neq 1 \qquad C = \text{const.},$$

where $\theta(x-1) = 1$ for $x > 1$ and zero otherwise. Both f_1 and f_2 coincide with $f(x)$ in $-1 < x < 1$. Both f_1 and f_2 are infinitely differentiable except at $x = 1$. Yet they differ for $x > 1$. This is so because there is no way to "continue" the function $f(x)$ through its singular point $x = 1$ for a single real variable. On the other hand, were one dealing with a complex variable z, one could pass "around" an isolated singular point because one is now dealing with a complex plane. Thus if we consider the function

$$f(z) = 1 + z + z^2 + \ldots \tag{8.1}$$

it is known to be analytic in $|z| < 1$. Now, we will see that there is a *unique* function

$$f_1(z) = \frac{1}{1-z} \qquad z \neq 1 \tag{8.2}$$

which coincides with $f(z)$ in $|z| < 1$ and is analytic in a wider domain (in fact in the largest domain possible, i.e., complex plane minus the point $z = 1$).

Now, we shall elaborate on the concept of analytic continuation further.

Suppose, one knows an analytic function, but it is defined over a limited domain D_0. For example, consider

(1) $f(z) = \sum_{n=0}^{\infty} z^n$ defined only for $|z| < 1$ and is analytic there.

(2) $\Gamma(z) = \int_0^{\infty} e^{-t} t^{z-1} dt$ is defined for Re $z > 0$, and is analytic in this domain.

Now, one is interested in knowing, in each case, whether an analytic function can be defined over a wider domain $D \supset D_0$ such that it coincides with the above function in D_0. Typically, one would proceed as follows: Choose a point $z = z_0$, say near the boundary of the domain of definition D_0 of the function $f(z)$. Here the function $f(z)$ is analytic, and hence all the derivatives exist at $z = z_0$. One can then expand $f(z)$ in a Taylor series around $z = z_0$. The Taylor series will converge in a circle of radius, say, R, around $z = z_0$. In this circle, then, the Taylor series represents an analytic function. Now, it is possible that this circle *may* contain points outside the original domain of definition. Let $D_1 = D_0 \cup \{|z - z_0| < R\}$. We, now, have a single analytic function defined over a larger domain D_1. Thus, the domain of definition of the analytic function has been enlarged. In other words, we now say that the function $f(z)$ is analytically continued from D_0 to a larger domain D_1. And, this process could conceivably be continued to extend the original domain of definition D_0 to as large a domain as possible (This will be discussed in more detail later.)

To understand the above process concretely, let us consider an example. Let

$$f(z) = \sum_{n=0}^{\infty} z^n \qquad |z| < 1. \tag{8.3}$$

In this case, of course, we already know the analytic continuation (given earlier). But we shall treat this case as an illustrative example to see how analytic continuation may be carried out for a general analytic function. So we will, for the present, not assume the known result for its analytic continuation.

Here, the domain of definition of f, viz. D_0 is $\{z : |z| < 1\}$. We could choose any z_0 in D_0 to construct the Taylor series around it, and z_0 near the boundary of D_0 (away from $z = 1$) is expected be more beneficial to enlarge the domain of definition of f. But for calculational simplicity, we shall choose $z_0 = -\frac{1}{2}$.

Now, to construct the Taylor series, we need to calculate $f^{(n)}(z_0)$ from (8.3). However, in *this* case, we may alternately do so more conveniently if we recognize its compact form $f(z) = \frac{1}{1-z}$; $|z| < 1$.

Then,

$$f^{(n)}(z) = \frac{n!}{(1-z)^{n+1}}.$$

Thus,

$$f^{(n)}(z_0) = \frac{n!}{\left(1 + \frac{1}{2}\right)^{n+1}} = n! \left(\frac{2}{3}\right)^{n+1}$$

Thus the Taylor series about $z = z_0$ becomes

$$f(z) = \sum_{n=0}^{\infty} n! \left(\frac{2}{3}\right)^{n+1} \frac{\left(z+\frac{1}{2}\right)^n}{n!}$$
$$= \frac{2}{3}\sum_{n=0}^{\infty}\left[\frac{2}{3}\left(z+\frac{1}{2}\right)\right]^n. \tag{8.4}$$

This series clearly converges for $\left|\frac{2}{3}\left(z+\frac{1}{2}\right)\right| < 1$, i.e., for $\left|z+\frac{1}{2}\right| < \frac{3}{2}$. We shall define $D_1 \equiv \left\{z : |z+\frac{1}{2}| < \frac{3}{2}\right\}$ and this domain is shown in Fig. 8.1. Note that D_1 contains D_0. Thus, the domain of definition of function $f(z)$ has been extended from D_0 to a larger domain D_1. Or alternately, we say that the function $f(z)$ has been analytically continued from D_0 to D_1.

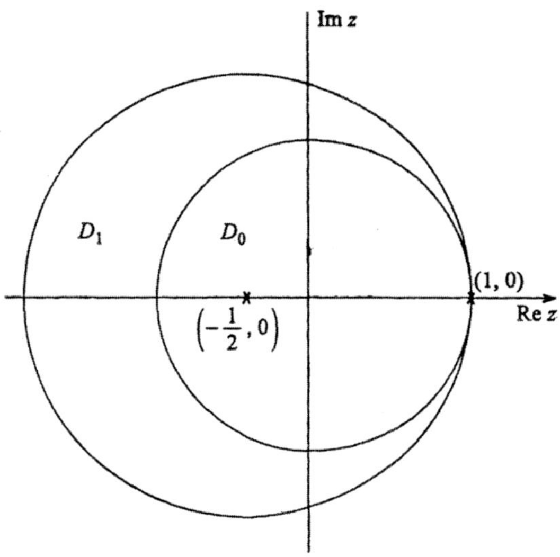

Fig. 8.1 *Analytic continuation for D_0 to D_1.*

Definition: Let $f_1(z)$ be analytic in domain D_1. Consider a domain D_2 with $D_1 \cap D_2 \neq \phi$ and $D_2 \not\subset D_1$ (D_2 overlapping with D_1 and extending beyond D_1). If we can find a function $f_2(z)$ in D_2 such that (i) $f_2(z)$ is analytic in D_2; and (ii) $f_1(z) = f_2(z)$ whenever $z \in D_1 \cap D_2$ (i.e., in the overlap region), then $f_2(z)$ is called the analytic continuation of $f_1(z)$ to the domain D_2.

The analytic function is now defined over an enlarged domain $D_1 \cup D_2$.

The process outlined in the previous example can be extended further and further. For example, one could take the point $z_1 = -\frac{3}{2}$ which is contained in D_1 and where $f(z)$ is analytic so that we could construct the Taylor series (at least in principle) by evaluating all order derivatives $f^{(n)}(z_1)$. For this the analogue of the series form (8.4) that defines $f(z)$ in

D_1 will have to be used in general case (even though the form (8.2) happens to be available in this simple setting). We will then find that now the Taylor series so constructed converges in the still larger circle $|z + \frac{3}{2}| < \frac{5}{2}$. Let us call this the domain D_2. Then $f(z)$ has been analytically continued successively from D_0 to D_1 to D_2. (See Problem 8.1.) In this case the process can be continued indefinitely and one can construct an analytic function $f(z)$ in the entire complex plane (except $z = 1$) and, of course, we can show explicitly that the result will be equivalent to

$$f(z) = \frac{1}{1-z} \qquad z \neq 1.$$

This is also expected from the uniqueness results in the next subsection.

8.1.1 Uniqueness of analytic continuation

At the end of the previous section we outlined a procedure that could be used to continue analytically a given function $f(z)$ to a wider domain. The procedure is however not unique because at each stage the choice of point around which a new Taylor expansion is made is not unique. A natural question arises that if a function $f_1(z)$, analytic over D_1, is analytically continued to $D_2(D_2 \cap D_1 \neq \phi)$, would one get a unique result for the analytic continuation irrespective of the procedure.

In this connection, we will firstly state a result, see its consequences and then indicate how we could go about proving it.

Theorem 8.1 Let $f_1(z)$ and $f_2(z)$ be two functions analytic within region D. If the two functions coincide (i) in a domain $D_1 \subset D$ (including, in particular, a neighborhood of point $z \in D$), or (ii) on an arc in D or (iii) on an infinite set of points with a limit point in D; then they coincide throughout D.

Before indicating the sum and substance of the proof, let us see two of its obvious implications. Consider two overlapping domains D_1 and D_2 with $D_1 \cap D_2 = D_3 \neq \phi$. Let $f_1(z)$ be analytic over D_1. Suppose there is a function $f_2(z)$ analytic over D_2 and coincident with $f_1(z)$ over the overlap $D_1 \cap D_2 = D_3$. Then $f_2(z)$ would be an analytic continuation of $f_1(z)$ over to D_2. We ask the question: Is $f_2(z)$ the unique analytic continuation? To see that this is so, let, if possible, there be another function $\tilde{f}_2(z)$ analytic in D_2 such that $\tilde{f}_2(z) = f_1(z)$ over D_3. Now consider the two functions $F(z)$ and $\tilde{F}(z)$ defined over $D_1 \cup D_2$.

$$\left.\begin{aligned} F(z) &= f_1(z) \qquad z \in D_1 \\ &= f_2(z) \qquad z \in D_2 \end{aligned}\right\}$$

and

$$\left.\begin{aligned} \tilde{F}(z) &= f_1(z) \qquad z \in D_1 \\ &= \tilde{f}_2(z) \qquad z \in D_2 \end{aligned}\right\}$$

Then these functions F and $\tilde{F}$ coincide over $D_3 = D_1 \cap D_2$ and are analytic in $D = D_1 \cup D_2$. Hence by the preceding theorem

$$F(z) = \tilde{F}(z) \qquad z \in D$$

In particular, therefore,

$$f_2(z) = \tilde{f}_2(z) \qquad z \in D_2$$

establishing the uniqueness of analytic continuation $f_2(z)$ over the domain D_2.

The uniqueness theorem further provides one with a procedure for establishing analytic continuations in an easy manner for many functions. Suppose one has a function $f_1(z)$ analytic over D_1 and suppose one can *guess* a function $f(z)$ analytic over a bigger domain $D(D_1 \subset D)$ such that $f(z) = f_1(z)$ over D_1. Then, from the uniqueness theorem, it must be *the* analytic continuation. There is then no need to follow any laborious continuation procedure to obtain $f(z)$ from $f_1(z)$.

Examples

(1) Here we simply take again the example which we have already studied in depth, but look at it from a new angle.
Let

$$f_1(z) = \sum_{n=0}^{\infty} z^n \qquad |z| < 1$$

defined over a domain D_1, the interior of a unit circle. Now we know that the function

$$f(z) = \frac{1}{1-z} \qquad z \neq 1$$

defined over the domain D (which is the entire complex plane with $z = 1$ omitted) coincides with $f_1(z)$ over D_1 and is analytic over D. Hence, *without following any procedure* for successive analytic continuations from D_1 to D, we *know* from uniqueness theorem that $f(z)$ defined above is *the* analytic continuation of $f_1(z)$ from D_1 to D.

(2) The uniqueness theorem allows one to conclude in the uniqueness of the function $f(z)$ analytic over D (if it exists), given the values of a function over an arc in D. Now suppose, we want a function analytic in entire complex plane that reduces to $f(x) = e^x$ over the real axis. Uniqueness theorem then implies that such a function, if it exists, is unique. Now we know that

$$f(z) = e^z$$

is entire (see Chapter 6) and does reduce to e^x on the real axis. It then is the only entire function with these properties.

Finally, we shall close the discussion by indicating the line along which the proof of uniqueness theorem proceeds. We do not propose to give the entire proof but follow only the underlying idea; the latter being simple compared to the elaborate mathematical construction required to realize it.

First, we reformulate the uniqueness theorem somewhat differently.

Theorem 8.2 If a function f is analytic throughout a domain D and vanishes (i) at each point of $D_1 \subset D$ or (ii) on an arc in D; then $f(z) \equiv 0$ throughout D. (We will not discuss the third possibility in the theorem here.)

We now outline the underlying idea of the proof. For concreteness, we shall consider the second possibility in which $f(z) = 0$ on an arc in D. We want to prove that $f(z)$ is zero at every point in D. So consider a point P in D and any point Q on the given arc. Since the domain is always a connected region, P and Q can be joined together by a path of finite length lying entirely *within* D. Let the distance of the closest approach of this path with the boundary of D be d. This is to say, every point on the path (entirely within D) is at a distance 'd' or greater from the boundary of D. Now as the function f is analytic in D, it can be Taylor expanded around Q. The various derivatives of f at Q, which can be evaluated along any direction for an analytic function, can in particular be evaluated along this arc. Since $f(z)$ is zero along the arc, all its derivatives at Q also vanish. Hence each term in the entire Taylor expression of $f(z)$ vanishes. Thus the analytic function $f(z)$, which is uniquely given by its Taylor expansion within the circle of convergence of the series, must be vanishing in this circle. But the radius of the circle R must be at least equal to d, or else it would imply the existence of a singularity at a distance less than d from Q (i.e., within D) which is impossible since $f(z)$ is analytic in D. Thus $f(z)$ is now known to vanish in a circle of radius $R \geq d$. We repeat this exercise now by taking a point in this circle near the circumference and carrying out the Taylor expansion. This again must converge in a circle of radius d (or greater) which now extends beyond the previous circle. This process, outlined in § 8.1 can be repeated till we reach the point P. The only question that needs to be considered is whether this process can be carried out in a finite number of steps. Using the two facts that the arc joining P and Q has a finite length and all the circles of convergence have a minimum radius of convergence d, this can be shown by an elaborate but straightforward mathematical construction. In this fashion $f(z)$ can be shown to vanish at the point P within the domain D and this is valid for an arbitrary P. Hence, etc.

Corollary: If a function $F(z)$ is analytic throughout a domain D and its value is known (i) at each point of $D_1 \subset D$, or (ii) on an arc in D, then $F(z)$ is known (uniquely) throughout D.

Proof: Suppose, if possible, there are two functions $f_1(z)$ and $f_2(z)$, both analytic within D, such that both coincide with F within D_1 (or on an arc), then $f(z) = f_1(z) - f_2(z)$ satisfies the conditions of the previous theorem. Hence $f_1 - f_2 \equiv 0$ throughout the domain D. Hence the function $F(z)$ is uniquely known.

8.2 Conformal Mapping and its Applications to Electrostatics

Many electrostatics problems in two dimensions are easily solved by the use of conformal mapping. By a two-dimensional problem (in the context of a realistic three-dimensional situation) we mean a problem in which the electrostatic potential ϕ is independent of one of the (cartesian) variables, say z; thus $\phi = \phi(x, y)$. This can happen on account of the translational symmetry of the physical situation along the z-axis. Since ϕ is now a function of only two variables x and y, we can, instead of the real variables x and y, introduce the complex variable $z = x + iy$. It is thus that the theory of functions of a complex variable becomes relevant in electrostatics. To be more specific, the use of the theory of analytic functions in the context of 2-D electrostatics arises from the observation that the real and the imaginary parts of an analytic function $f(z) = u(x, y) + iv(x, y)$ satisfy the Laplace's equation in two dimensions:

$$\left[\frac{\partial^2}{\partial x^2} + \frac{\partial^2}{\partial y^2}\right] u(x, y) = 0 = \left[\frac{\partial^2}{\partial x^2} + \frac{\partial^2}{\partial y^2}\right] v(x, y). \tag{8.5}$$

This suggests that we may be able to construct an analytic function $f(z)$ whose real (or imaginary) part is identified with $\phi(x, y)$, which itself staisfies Laplace's equation in free space. Now one could conceive of performing an analytic transformation $z = z(w)$ on f; then $f[z(w)] = h(w)$ will be analytic in w in a certain domain in the complex w plane. The real and the imaginary parts of $h(w)$, viz. $U(X, Y)$ and $V(X, Y)$ (here $w = X + iY$) satisfy the harmonic condition in the (X, Y) plane:

$$\left[\frac{\partial^2}{\partial x^2} + \frac{\partial^2}{\partial Y^2}\right] U(X, Y) = 0 = \left[\frac{\partial^2}{\partial X^2} + \frac{\partial^2}{\partial Y^2}\right] V(X, Y). \tag{8.6}$$

Any transformation such as $z = z(w)$ then allows one to generate solutions to the Laplace's equation in some domain and hence correspond to electrostatic potetial of *some* (new) electrostatic problem. Thus knowing one solution, $\phi(x, y)$, of a certain electrostatic problem, one is able to generate solutions to an infinite variety of new electrostatic problems (of which some may be of interest in practice.) This is the principle behind the use of analytic mapping in the 2-D electrostatic problems. We now explain in detail the above statements and illustrate with examples.

8.2.1 Construction of an analytic function

The electrostatic potential $\phi(\mathbf{r})$ in free space satisfies Laplace's equation

$$\nabla^2\phi = \left[\frac{\partial^2}{\partial x^2} + \frac{\partial^2}{\partial y^2} + \frac{\partial^2}{\partial z^2}\right] \phi(x, y, z) = 0.$$

Suppose an electrostatic problem has invariance under translations in a given direction. Then choosing this direction to be z-axis, the potential ϕ must satisfy

$$\phi(x, y, z) = \phi(x, y, z + a) \qquad a \text{ being arbitrary.}$$

This then requires

$$\frac{\partial \phi(\mathbf{r})}{\partial z} \equiv 0 \qquad \text{i.e., } \phi = \phi(x, y). \tag{8.7}$$

and ϕ then satisfies:

$$\left[\frac{\partial^2}{\partial x^2} + \frac{\partial^2}{\partial y^2}\right] \phi(x, y) = 0. \tag{8.8}$$

Now suppose a 2-D electrostatics problem is given so that (8.8) is satisfied in a domain D in the x–y plane. Then the potential $\phi(x, y)$ is uniquely fixed once apporpriate boundary conditions (say Dirichlet or Neumann) on the boundary S of D are given.

We now begin with a specific solution ϕ of (8.8) over a particular domain D and with specific boundary conditions on S. We will then construct a function $f(z)$, analytic over D, such that $\phi(x, y)$ is its real part.

Consider a function

$$f(z) = \phi(x, y) + i v(x, y) \tag{8.9}$$

where $v(x, y)$ is yet to be dertermined. We want $f(z)$ to be analytic; so, in particular (ϕ, v) must satisfy the CR conditions. Thus

$$\frac{\partial v}{\partial y} = \frac{\partial \phi}{\partial x} = g(x, y); \text{ a known function over } D \tag{8.10}$$

$$-\frac{\partial v}{\partial x} = \frac{\partial \phi}{\partial y} = h(x, y); \text{ a known function over } D \tag{8.11}$$

Thus,

$$v(x, y) = \int^{y} g(x, y')dy' + C(x), \tag{8.12}$$

where the integral is carried out keeping x constant. To determine $C(x)$, we use (8.11),

$$\begin{aligned}\frac{\partial v}{\partial x} &= \frac{\partial}{\partial x}\int^{y} g(x, y')dy' + C'(x) = \int^{y} \frac{\partial g}{\partial x}(x, y')dy' + C'(x)\\ &= \int^{y} \frac{\partial^2 \phi}{\partial x^2}(x, y')dy' + C'(x)\end{aligned}$$

$$= -\int^{y} \frac{\partial^2 \phi}{\partial y'^2}(x, y')dy' + C'(x) \quad \text{on account of (8.8)}$$

$$= -\frac{\partial \phi}{\partial y}(x, y) + C'(x). \tag{8.13}$$

Comparing (8.13) with (8.11) we find that

$$C'(x) = 0 \qquad \text{i.e., } C = \text{const.} \tag{8.14}$$

Thus

$$v(x, y) = \int^{y} g(x, y')dy' + C. \tag{8.15}$$

This determines v upto a constant.

Of course, CR conditions are not sufficient for proving the analyticity of $f(z)$. We also need the existence and continuity of first order partial derivatives of ϕ and v. But this is guarenteed in a physical context. To see this, we note that $-\frac{\partial \phi}{\partial x}$ and $-\frac{\partial \phi}{\partial y}$ are components of the electric field in D and hence expected to be finite (note the absence of point and line charges in D). In a region free of charges (free and bound) we expect the electric field to be a continuous function. (This is seen by noting that the electric field in this region is being produced by superposition of electric fields created by charge elements outside D and these from Coulomb's law are continuous functions of position). Thus from (8.10) and (8.11) $\frac{\partial v}{\partial y}$ and $\frac{\partial v}{\partial x}$ are therefore continuous throughout D. Hence, the function $f(z)$ of (8.9) so constructed is analytic within D. The constant C in (8.15) can be dropped as we are interested only in one analytic function whose real part is the potential function. Note, alternately, that we could also have constructed an analytic function $\tilde{f}(z)$ whose imaginary part coincides with $\phi(x, y)$ (as it is also harmonic). In fact it is easy to see that the function

$$\tilde{f}(z) = if(z) = -v(x, y) + i\phi(x, y) \tag{8.16}$$

is such an analytic function (modulo a constant).

Example 1 Consider the electric field $\mathbf{E} = E_0\hat{\mathbf{i}} =$ constant in the domain D which is the right half plane $x > 0$. It is described by the potential function

$$\phi(x, y) = -E_0 x + C \tag{8.17}$$

which clearly satisfies the Laplace's equation (8.8) in D. We define $v(x, y)$ by

$$\frac{\partial v}{\partial x} = -\frac{\partial \phi}{\partial y} = 0 \qquad \frac{\partial v}{\partial y} = \frac{\partial \phi}{\partial x} = -E_0 \tag{8.18}$$

which is solved as

$$v(x, y) = -E_0 y + C' \tag{8.19}$$

leading to

$$f(z) = -E_0 z + (C + iC'). \tag{8.20}$$

The constant C' could be dropped for reasons explained in the earlier paragraph. The constant C in (8.17) could also be dropped because ϕ is always ambiguous upto a constant and needs a convention to fix it uniquely. We could then work with the simple analytic function

$$f(z) = -E_0 z. \tag{8.21}$$

Example 2 Let $E_x = ay$; $E_y = ax$ and let the domain D be the first quadrant ($x > 0$, $y > 0$). The potential function (with suitable convention) can be taken as

$$\phi(x, y) = -axy.$$

We now define

$$f(z) \equiv \phi(x, y) + iv(x, y)$$

with $v(x, y)$ satisfying (8.10) and (8.11) (the imposed CR conditions)

$$\frac{\partial v}{\partial x} = -\frac{\partial \phi}{\partial y} = ax, \tag{8.22}$$

$$\frac{\partial v}{\partial y} = -\frac{\partial \phi}{\partial x} = -ay. \tag{8.23}$$

Integrating (8.22), we obtain

$$v = \frac{1}{2}ax^2 + C(y). \tag{8.24}$$

Then (8.23) constraints the v of (8.24):

$$C'(y) = -ay \tag{8.25}$$

so that

$$C(y) = -\frac{1}{2}ay^2 \tag{8.26}$$

where we have dropped the obvious constant in (8.26). This leads to

$$f(z) = -axy + i\frac{1}{2}a(x^2 - y^2) \equiv \frac{ia}{2}z^2 \tag{8.27}$$

an entire function.

Comment: We found out the analytic functions $f(z)$ whose real part is the electrostatic potential $\phi(x, y)$ for a certain electrostatic problem. Now if we consider a related analytic function $\tilde{f}(z) = if(z)$, its real part $v(x, y)$ is also harmonic and hence represents the electrostatic potential of some other electrostatic problem in the same domain.

For example, in Example (1)

$$\tilde{f}_1(z) = -if_1(z) = iE_0 z = -E_0 y + iE_0 x,$$

and the real part $\equiv \tilde{\phi}$ is

$$\tilde{\phi}_1(x, y) = -E_0 y$$

and represents a uniform electric field in the y-direction in the right half plane.

In Example (2),

$$\tilde{f}_2(z) \equiv -if_2(z) = \frac{a}{2}z^2$$

whose real part $\tilde{\phi}_2$ is

$$\tilde{\phi}_2(x, y) = \frac{a}{2}(x^2 - y^2).$$

This represents the electric field configuration

$$E_x = -ax \qquad E_y = ay$$

in the first quadrant.

8.2.2 Analytic transformation and electrostatic problems

We now utilize the simple examples in the previous subsection to illustrate the use of analytic transformations in electrostatic problems.

Consider, for example, the electrostatics problem in Example (1) of Section 8.2.1. We have

$$\phi(x, y) = -E_0 x \qquad x > 0 \tag{8.28}$$

representing a uniform electric field $\mathbf{E} = E_0\hat{\mathbf{i}}$ in the half plane $x > 0$. This was associated with an analytic function

$$f(z) = -E_0 z. \tag{8.29}$$

Now imagine performing an analytic transformation

$$z = w^2 \tag{8.30}$$

with

$$w = X + iY. \tag{8.31}$$

We then obtain a new function

$$f(z) \equiv h(w) = -E_0 w^2. \tag{8.32}$$

We note that, in view of the fact that

$$\arg z = 2 \arg w = 2 \tan^{-1} \frac{Y}{X} \tag{8.33}$$

and that in the domain D with $x > 0$, we have

$$-\pi/2 < \arg z < \pi/2, \tag{8.34}$$

the corresponding domain D' in complex w-plane is given by

$$-\pi/4 < \arg w < \pi/4. \tag{8.35}$$

The domain D is mapped on to D' (with one-one correspondence) by (8.30).

As the transformation $z = w^2$ is an analytic transformation of the function $f(z)$ of (8.29), the function so obtained, viz. $h(w)$ of (8.32) is also an analytic function of w in its domain D'. Thus the real (and also the imaginary part) of the function $h(w)$ must be harmonic and hence must represent solution to some (other) electrostatic problem. We note

$$h(w) = -E_0 w^2 = -E_0(X^2 - Y^2) - 2iE_0 XY \tag{8.36}$$

$$\equiv \Phi(X, Y) + iV(X, Y). \tag{8.37}$$

As already noted in the Example (8.4) of the previous subsection,

$$\Phi(X, Y) = -E_0(X^2 - Y^2) \tag{8.38}$$

is harmonic in X and Y. It thus represents a solution to the electrostatic problem in the domain D' (defined by ($|\arg w| < \pi/4$, i.e., $X > |Y|$) with the boundary condition $\Phi \equiv 0$ on the boundary $X = |Y|$.

We thus note that starting from one simple electrostatic problem, we have generated (by an analytic transformation) a solution to quite a different and somewhat more complicated electrostatic problem. But one could conceive of performing any (of the infinite) analytic transformation; each then leading to the solution for a new electrostatic problem.

8.2.3 Conformal mapping

Having seen some of the examples of application of analytic tranformations, we now go on to show an important property of analytic transformations that gives it the name of *conformal* transformations.

Theorem 8.3 Let $w = f(z)$ be analytic over domain D_z. Let C_1 and C_2 be any two smooth arcs in D_z intersecting at $z_0 \in D_z$. Let D_z be mapped onto D_w in the complex w-plane with C_1 and C_2 being mapped onto C_1' and C_2', and let $w_0 = f(z_0)$. Let θ_{12} and θ_{12}' be the angles between tangents to C_2 and C_1 and to C_2' and C_1' respectively. Then, provided $f'(z_0) \neq 0$,

$$\theta_{12} = \theta_{12}',$$

i.e., analytic transformations are angle-preserving or *conformal.*

Proof: Let C_1 and C_2 be parameterized by

$$z = z_1(t_1) \quad \text{and} \quad z = z_2(t_2) \tag{8.39}$$

respectively. C_1' and C_2' can be parameterized by

$$w = f(z_1(t_1)) \equiv w_1(t_1) \text{ and } w = f(z_2(t_2)) = w_2(t_2). \tag{8.40}$$

Now the tangent to C_1 makes an angle arg $\left[\dfrac{dz_1}{dt_1}\right]$ with the x-axis and thus

$$\theta_{12} = \left\{\arg\frac{dz_2}{dt_2} - \arg\frac{dz_1}{dt_1}\right\}\bigg|_{z=z_0} \tag{8.41}$$

In an identical manner

$$\theta_{12}' = \left\{\arg\frac{dw_2}{dt_2} - \arg\frac{dw_1}{dt_1}\right\}\bigg|_{w=w_0} \tag{8.42}$$

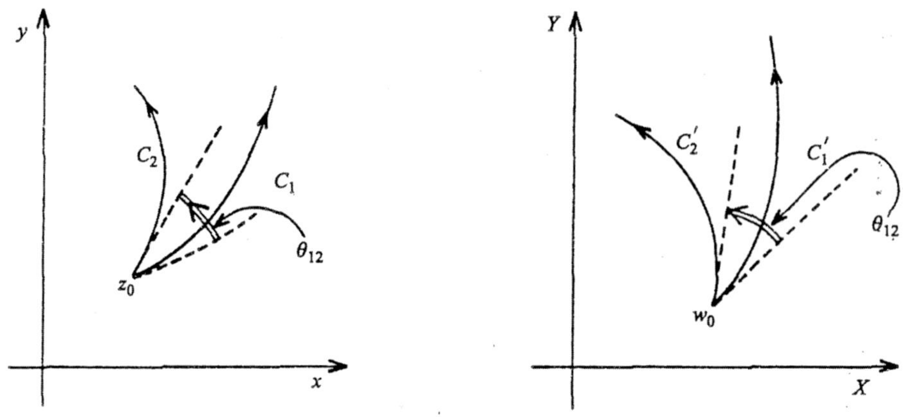

Fig. 8.2 *Conformal mapping*

but

$$\frac{dw_1}{dt_1} = \frac{d}{dt_1} f(z_1(t_1)) = \left[\frac{df}{dz} \cdot \frac{dz_1(t_1)}{dt_1}\right]\Bigg|_{z_0}. \tag{8.43}$$

Thus

$$\arg \frac{dw_1}{dt_1} = \arg \frac{df}{dz}\Bigg|_{z=z_0} + \arg \frac{dz_1(t_1)}{dt_1}\Bigg]\Bigg|_{z_0}. \tag{8.44}$$

In a similar manner,

$$\arg \frac{dw_2}{dt_2} = \arg \frac{df}{dz}\Bigg|_{z=z_0} + \arg \frac{dz_2(t_2)}{dt_2}\Bigg]\Bigg|_{z_0}. \tag{8.45}$$

From (8.44), (8.45), (8.41) and (8.42), the result

$$\theta_{12} = \theta'_{12}$$

follows.

8.3 Method of Steepest Descent

We shall introduce a method for an approximate evaluation of certain contour integrals. The method finds applications whenever contour integrals of a certain type (to be specified) are to be approximately evaluated. For example, many special functions (see the accompanying volume *Mathematical Physics: Advanced Topics*) of mathematical physics have contour integral representations. Using the method of steepest descent, these can be approximately evaluated in certain limits (e.g. $\Gamma(n)$ function for large n, Bessel function $J_\nu(x)$ for large x, etc.).

We introduce the method by first considering a familiar real integral. We shall bring out a number of facets of the method with the help of it that are relevant in the general application of the method of steepest descent to contour integrals. Then we shall outline the general procedure. We shall give a few more complex examples of the method later on.

8.3.1 An example

Consider the real integral (n a +ve integer)

$$\int_0^\infty dt\, e^{-t}\, t^{n-1}\, dt = (n-1)! \equiv \Gamma(n). \tag{8.46}$$

The above function $\Gamma(n)$ can be generalized to complex arguments w (more on this in Chapter 2 of the accompanying volume *Mathematical Physics: Advanced Topics*) by

$$\Gamma(w+1) \equiv \int_0^\infty dt\, e^{-t}\, t^w dt \qquad \text{Re } w > -1 \tag{8.47}$$

(the condition $Re\, w > -1$ is needed for convergence of the integral near $t = 0$). This integral cannot in general be evaluated exactly unless $w = n$ or $(n + 1/2)$. Suppose we now consider the problem of an approximate evaluation of this integral. For simplicity we shall take w to be real (but not integral) and positive. We look at the integrand

$$f(t) = e^{-t}\, t^w$$

Over the interval $(0, \infty)$, e^{-t} is an exponentially decaying function whereas t^w is a rising function $(w > 0)$. The product generally shows the behaviour deplicted in Fig. 8.3.

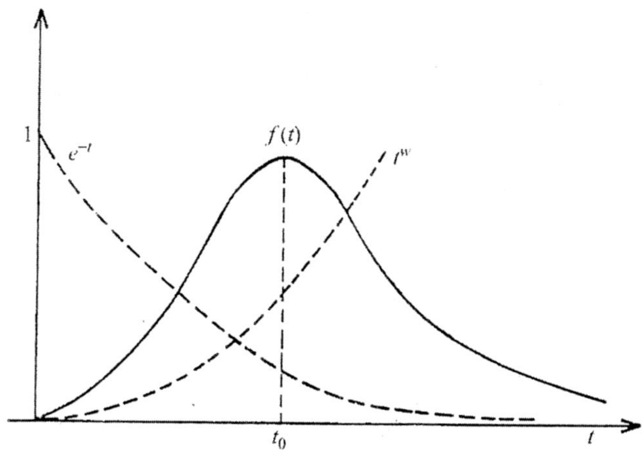

Fig. 8.3 *Graphs of* e^{-t}, t^w *and* $f(t) = e^{-t}t^w$

The function $f(t)$ has a single maximum at $t = t_0 = w$. The function is large in the neighborhood of $t = t_0$. The function falls off rapidly as $t \to \infty$ and for large w it also falls of rapidly as $t \to 0$. Thus for large w, there is a limited range around $t = t_0$ over which integrand is substantial (see Problem 8.11). Over this region and *near* $t = t_0$, the overall shape of the curve is approximately Gaussian. Thus for large w, we could think of evaluating the integral firstly by limiting the range of variable t; secondly by replacing the curve in the limited range by a Gaussian centered around $t = t_0$ (of appropriate width yet to be found) and thirdly by extending the Gaussian to $(-\infty, \infty)$ since the Gaussian is known to fall off rapidly at both ends. Now, how good the above procedure is depends on (i) how sharply the function $f(t)$ is peaked around $t = t_0$; (ii) how rapidly it falls to zero away from the peak and (iii) how good is the Gaussian approximation for the region near the peak. We shall now see that as $w \to \infty$, the accuracy improves rapidly and the method in fact gives an excellent approximation.

To understand the above points quantitatively, we first make a change of variables $t = wx$.

Then,

$$\begin{aligned}\Gamma(w+1) &= \int_0^{\infty} e^{-wx}(wx)^w\ w\ dx \\ &= w^{w+1}\int_0^{\infty} e^{-wx}x^w\ dx \\ &= w^{w+1}\int_0^{\infty} e^{-w(x-\ln x)}\ dx.\end{aligned} \tag{8.48}$$

Now we deal with the integrand of the form

$$f(x,w) \equiv e^{-w\ f(x)}$$

We note that $f(x)$ is minimum at $x = x_0$ given by

$$f'(x_0) = 0 = 1 - \frac{1}{x_0} \qquad \text{i.e. at } x_0 = 1.$$

Now the integrand can be written as

$$f(x,w) = e^{-wf(1)}\ e^{-w[f(x)-f(1)]} = e^{-w(x-1-\ln x)}$$

We thus have

$$\frac{\Gamma(w+1)}{e^{-w}w^{w+1}} = \int_0^{\infty} e^{-w(x-1-\ln x)}dx \equiv I(w). \tag{8.49}$$

Now to evaluate $I(w)$, we note the behaviour of the integrand around $x = 1$ for a large w such as $w = 1000$:

$$\begin{aligned} e^{-w(x-1-\ln\ x)} &= e^{-w(x-1)} = 1 && \text{at} \quad x = 1 \\ &\simeq 0.02 && \text{at} \quad x = 1.090 \\ &\simeq 0.02 && \text{at} \quad x = 0.9. \end{aligned}$$

Thus the integrand is sharply peaked in the region (0.9, 1.09) and is less than 2% of its peak value outside this integral. It should be noted that as this peak becomes sharper, the larger the w. (See Problem 8.12.) We can thus replace this integral by an integral over such an interval $1 - \xi_1, 1 + \xi_2$

$$\int_0^{\infty} e^{-w(x-1-\ln x)}\ dx \simeq \int_{1-\xi_1}^{1+\xi_2} e^{-w(x-1-\ln x)}\ dx. \tag{8.50}$$

Now this integral is by no means easy to evaluate analytically. We now wish to replace it approximately by a Gaussian integral that can indeed be evaluated analytically. To this end, we note the Taylor expansion of $(x - 1 - \ln x)$ around $x = 1$.

$$x - 1 - \ln x = 0 + 0 + \frac{1}{2}(x-1)^2 + O(x-1)^3. \tag{8.51}$$

In the above interval the $O(x-1)^3$ terms are small and can, to a good approximation (a *new* one), be ignored. We then have

$$I(w) = \int_{1-\xi_1}^{1+\xi_2} e^{-w\frac{1}{2}(x-1)^2}\,dx. \tag{8.52}$$

Now, again, the Gaussian itself is sharply peaked around $x = 1$. Thus for $w = 1000$,

$$\begin{aligned} e^{-\frac{w}{2}(x-1)^2} &= 1 & x &= 1 \\ &= 0.017 & x &= 1.09 \\ &= 0.007 & x &= 0.9. \end{aligned}$$

So we can replace $I(w)$ an integral over an infinite range, viz :

$$I(w) \simeq \int_{-\infty}^{\infty} e^{-w\frac{1}{2}(x-1)^2}\,dx$$

since outside the interval $(1 - \xi_1, 1 + \xi_2)$, the contribution to the integral is negligible. But the integral can be easily evaluated and we find

$$I(w) \sim \sqrt{\frac{2\pi}{w}}.$$

This gives us the leading approximation for $\Gamma(w+1)$ for large w:

$$\frac{\Gamma(w+1)}{e^{-w}w^{w+1}} \simeq \sqrt{\frac{2\pi}{w}}. \tag{8.53}$$

It can be shown that a further correction to (8.53) is of $O\left(\frac{1}{w^{3/2}}\right)$. Thus

$$\Gamma(w+1) = e^{-w}w^{w+\frac{1}{2}}\sqrt{2\pi}\left[1 + O\left(\frac{1}{w}\right)\right]. \tag{8.54}$$

[See Problem 8.13 to know how this term is evaluated].

We note several features of the above calculation:

(i) We were given a definite integral the integrand of which is large in a given region and negligible outside. The point around which it is large is, of course, found by locating its maximum.
(ii) The integral depends on a parameter w.
(iii) The region where the integrand in (8.49) is substantial becomes narrower, depending on the value of w (here large). Thus for large w, we can effectively omit the remaining integration range.
(iv) In the range of interest, we cannot generally do the integral exactly. However, the integral can effectively be replaced by a Gaussian integral which is known exactly.

8.3.2 General considerations in the method

We shall now note the manner in which a problem in can differ from the above simple example and still be amenable to the method to be described.

[A] Firstly, the integral to be evaluated may be a contour integral. On account of the Cauchy–Goursat theorem, the contour for the integral need not be unique; it can be deformed keeping the end points fixed. $\left(\int_{C_1} f(z)\,dz = \int_{C_2} f(z)\,dz\right.$ if $f(z)$ is analytic within the region enclosed by C_1 and C_2 and on C_1 and $\left.C_2\right)$. Thus, given a contour integral there may not exist a *narrow segment of* C from where the main contribution to integral comes. However, as C can be deformed, all that is required is that a new contour C' exists such that the integrand is substantial only over a narrow segment of C'.
[B] It is not necessary that there is only one such region where the integrand is large. It is sufficient if there are finitely many such regions which are well separated. Then the integral around each can be replaced by a separate Gaussian and be evaluated.
[C] Just as the contour may have to be deformed so as to pass through region(s) of large values of the integrand, its orientation may also have to be chosen so that the contribution over that region comes from as short a contour as possible. This point is further illustrated with the help of the following example.

Example Consider the evaluation of the following integral for large w

$$I(w) = \int_0^\infty dx e^{-[x^3-9x^2+(18-12i)x]w} \equiv \int_0^\infty dx\, e^{-wf(x)}.$$

We examine the behaviour of the modulus of the integrand along the x-axis. The modulus of the integrand is

$$|e^{-wf(x)}| = e^{-w\,\mathrm{Re}\,f(x)}$$

$$\mathrm{Re}\,f(x) = x^3 - 9x^2 + 18x = x(x-3)(x-6).$$

We have tabulated value of $e^{-w \text{ Re } x}$ for some x (see table below).

x	$f(x)$	$e^{-w \text{ Re } f(x)}$
0	0	1
1	10	e^{-10w}
2	8	e^{-8w}
3	0	1
4	-8	e^{8w}
5	-10	e^{10w}
6	0	1
7	28	e^{-28w}
∞	$-\infty$	0

We thus see that the integrand is large mainly in the region (3, 6) and for $x > 6$ it rapidly vanishes. One would expect that we could approximate the integrand $I(w)$ by only the contribution given below:

$$I(w) \sim \int_3^6 dx\, e^{-wf(x)}.$$

However, evlauation of this proves difficult for two reasons. Even though the integrand is large in this region, there is a destructive interference due to the rapidly oscillating phase factor

$$\exp\{-w \text{ Im } f(x)\} = \exp(12\, iwx)$$

leading to a large cancellations. Secondly, the form of the integrand is not such as can lead to an easy evaluation. We now proceed with what is called the saddle point method.

We generalize the integrand to the complex plane

$$e^{-w\, f(z)} \equiv e^{-w(z^3-9z^2+(18-12i)z)}.$$

The saddle points are given by

$$f'(z) = 0 = 3z^2 - 18z + (18 - 12i) = 0,$$

i.e.

$$z_1 = 1 - i; \qquad z_2 = 5 + i$$

are the two saddle points.

Now let us examine the behaviour of the integrand $e^{-wf(z)}$ near the two saddle points.

At $z = z_1$,

$$|e^{-wf(z)}| = e^{-4w}$$

and at $z = z_2$

$$|e^{-wf(z)}| = e^{4w}.$$

We then have to deform the contour to pass through the saddle points. Near the saddle points, the direction through each must be chosen correctly as in [C] above. And away from the saddle points, it must be chosen so that the main contribution to $I(w)$ comes from the segments of the contour in the vicinity of the two saddle points, which we know how to estimate by Gaussian approximation. It then remains to verify that the rest of the contour indeed contributes insignificantly.

In view of the fact that the integrand is far more dominant at the second saddle point, the main contribution will come from this saddle point. Hence we shall deal with this one in detail and leave the results for the other one as an exercise (see Problem 8.16).

Before we proceed with the algebra, we note the contributions to $I(w)$ from the dominant regions of the original contour (+ve real axis) and that of the deformed contour (not yet entirely fixed). As noted earlier, the dominant contribution to $I(w)$ comes from the interval (3, 6) and the here the integrand is large in magnitude but with rapid oscillations. This is being replaced by an integrand much smaller in magnitude near $z = z_2$ but with no destructive interference on the deformed contour.

We now discuss the orientation of the deformed contour through $z = z_2$. We write

$$\begin{aligned} e^{-wf(z)} &\simeq e^{-wf(z_2)} e^{-\frac{1}{2}wf''(z_2)\,(z-z_2)^2} \\ &= e^{4w}\, e^{58wi}\, e^{-\frac{1}{2}w(12+6i)(z-z_2)^2}. \end{aligned}$$

Now the direction of the contour through the saddle point is determined by the requirement that the imaginary part of the exponent is a constant in the neighborhood of $z = z_2$. We parameterize a small portion of the contour (taken as a straight line) by

$$(z - z_2) = \rho e^{i\phi}.$$

Then

$$wf(z) \simeq 4w + 58wi - \frac{1}{2}(12 + 6i)\rho^2\, e^{2i\phi}\, w,$$

$$\text{Im}\,[wf(z)] = 58w - 6\rho^2 w \sin 2\phi - 3\rho^2 w\, \cos 2\phi.$$

$\text{Im} f(z)$ should be a constant as ρ (i.e., distance of z from z_2) is varied near $z = z_2$. This requires that

$$6w \sin 2\phi + 3w \cos 2\phi = 0,$$

$$\tan 2\phi = -\frac{1}{2}.$$

This has four independent solutions:

$$\phi = \frac{1}{2}\text{Tan}^{-1}\left(-\frac{1}{2}\right), \quad \frac{1}{2}\text{Tan}^{-1}\left(-\frac{1}{2}\right) + \frac{\pi}{2}, \quad \frac{1}{2}\text{Tan}^{-1}\left(-\frac{1}{2}\right) + \pi,$$
$$\frac{1}{2}\text{Tan}^{-1}\left(-\frac{1}{2}\right) + \frac{3\pi}{2}$$

Of these the first and the third solution represents the two sides of a single straight segment and so do the second and the fourth. Of these, the second and fourth solution represents the curve of the steepest ascent, i.e., the curve along which the $-\text{Re}[f(z)w]$ attains its minimum at z_2. To see this consider

$$\begin{aligned} -\text{Re}\,[f(z)w] &= 4w - \frac{1}{2}(12+6i)\rho^2 w\left[-\frac{2}{\sqrt{5}} + i\frac{1}{\sqrt{5}}\right] \\ &= 4w + 3(2+i)(+2-i)\rho^2 w/\sqrt{5} \\ &= 4w + \frac{9}{\sqrt{5}}\rho^2 w. \end{aligned}$$

Thus, along this direction, $|e^{-wf(z)}|$ is increasing as one goes away from $z = z_2 (\rho > 0)$. This line of steepest ascent is to be discarded. We instead take the first (and third) solution for which $|e^{-wf(z)}|$ is a maximum at $z = z_2$, and is (approximately) diminishing like a Gaussian on both sides.

In fact, if this fact is realized, the above discussion can be avoided. We required directly that

$$-\frac{1}{2}wf''(z_2)\rho^2 e^{2i\phi}$$

is real and negative. This requires that $\frac{1}{2}w(12+6i)\rho^2 e^{2i\phi}$ is real and positive.

Hence $e^{2\phi i}$ must be proportional to the complex conjugate of $(12+6i)$. Hence as $|e^{2i\phi}| = 1$,

$$e^{2i\phi} = \left(\frac{2-i}{\sqrt{5}}\right).$$

This gives the first and the third solution.

A similar discussion can be given for the first saddle point and the sense of the contour through $z = z_1$ can be determined (see Problem 8.16). The contributions from the two saddle points can be determined by approximating each by a Gaussian. The contribution from the rest of the contour has to be shown to be negligible. This is left as exercise. However, as emphasized earlier, the contribution from the second saddle point is far greater than the contribution from the first and hence the latter can be ignored. The result for the contribution from the second saddle point is

$$I(w) \simeq \sqrt{\frac{\pi}{3\sqrt{5}}}\, w^{-1/2}\, e^{4w}\, e^{58wi}.$$

8.3.3 General theory of the method of steepest descent

Firstly we shall restrict ourselves to an integral of the kind

$$J(\tau) = \int_C e^{\tau f(z)} dz,$$

where τ is a free real parameter and $f(z)$ is analytic in the domain D which includes C. We are required to obtain a good approximation to $J(\tau)$ for large τ. We ask if it is possible to deform the contour C (within the domain D) such that the integrand is large only over a small segment of the new contour C'. We note that

$$J(\tau) = \int_C e^{\tau(\mathrm{Re}\, f + i \mathrm{Im}\, f)}\, dz$$

$$\equiv \int_{C'} e^{\tau u(x,y)} e^{iv(x,y)}\, dz.$$

Now suppose we could deform the contour C to C' (if necessary) such that

(a) the modulus of the integrand $\exp[\tau u(x, y)]$ is large only in a small portion of C';
(b) the phase of the integrand $v(x, y)\tau$ is constant along C' *in this region* so that there is no destructive interference between various contributions to J;
(c) the integrand is negligible outside this portion of C';

then we could evaluate $J(\tau)$ by replacing C by C' and approximating the integral by one over this portion only. Further, the integral over this portion is approximately replaced by a Gaussian integral.

Here we note that condition (b), which chooses the *direction* in which C' passes through this region, ensures that the contribution comes from as small a segment of C' as possible, since it avoids destructive interference.

Now we formulate mathematically the conditions on locating such region(s). Consider a contour C' passing through a point P and so oriented that (i) u is maximum at P along C'; (ii) v is constant along C' in the vicinity of P. Then, at least the conditions (a) and (b) above will be fulfilled. Let $\hat{t}$ be the unit tangent vector to the curve at P. Thus (i) implies

$$\left.\frac{\partial u}{\partial s}\right|_{\mathrm{P}} = 0 = \nabla u \cdot \hat{\mathbf{t}}. \tag{8.55}$$

Condition (ii) further implies

$$\nabla v \cdot \hat{t} = \left.\frac{\partial v}{\partial s}\right|_{\mathrm{P}} = 0. \tag{8.56}$$

Now suppose, if possible, ∇u and ∇v are both nonzero. Then as ∇u and ∇v are orthogonal ($\nabla u \cdot \nabla v = 0$, see Chapter 6), (i) and (ii) would imply the existence of three mutually orthogonal vectors in x–y plane, which is impossible. Hence at least one of these must be zero. Let us suppose $\nabla u\Big|_{\mathrm{P}} = 0$, i.e., $\left.\frac{\partial u}{\partial x}\right|_{\mathrm{P}} = 0 = \left.\frac{\partial u}{\partial y}\right|_{\mathrm{P}}$. Then CR condition for $f(z)$ imply

$$\left.\frac{\partial v}{\partial x}\right|_{\mathrm{P}} = -\left.\frac{\partial u}{\partial y}\right|_{\mathrm{P}} = 0; \left.\frac{\partial v}{\partial y}\right|_{\mathrm{P}} = \left.\frac{\partial u}{\partial x}\right|_{\mathrm{P}} = 0,$$

i.e., $\nabla v \equiv 0$ also. This leads us further to

$$f'(z)\Big|_P = \left.\frac{\partial u}{\partial x}\right|_P - i\left.\frac{\partial u}{\partial y}\right|_P = 0.$$

So the region under consideration is located by finding where $f'(z)$ vanishes. We must choose C' to pass through this region in such a sense that the imaginary part v is stationary along C' in the vicinity of P.

Solutions to $f'(z) = 0$ are called the "saddle points" of $f(z)$ (for reasons explained later). If there are more than one saddle point, C' has to be chosen so as to go through each of these with a proper orientation. (It may however turn out that only one or a few of the saddle points gives dominant contribution. See Section 8.3.2). Further, the point (c) above that the rest of the contour C' does not contribute significantly, has to be ascertained in each case *explicitly.*

Finally we explain the two names given to this method : (i) the steepest descent method, (ii) the saddle point method. The contour C' is chosen to pass through P along such a direction that $\nabla v \cdot \hat{t} = 0$ along it. Then ∇u (which is also perpendicular to ∇v) must be parallel to $\hat{t}$, the sense of C'. But direction of ∇u tells us the direction in which u varies (here decreases) fastest. Thus along C', the modulus of the integrand $\exp[\tau u(x, y)]$ falls off fastest; or has its "steepest descent", as one goes away from P. Finally, we comment on the nature of the point P. At this point $\frac{\partial u}{\partial x} = 0 = \frac{\partial u}{\partial y}$ so that u is an extremum. But u, beirg the real part of an analytic function, satisfies

$$\frac{\partial^2 u}{\partial x^2} + \frac{\partial^2 u}{\partial y^2} = 0$$

and hence has no absolute maximum or a minimum (see Chapter 6). Thus as one moves along different curves through P, one may find that P is a point of maximum of u or a minimum of u or a point of inflexion for u along a given curve. Thus the graph of the modulus $\exp[\tau u(x, y)]$ around point P is shaped in the form of the "horse saddle" and P is a "saddle point" of $f(z)$.

We finally note that the above process can as well as applied to an integral of the form

$$J(\tau) = \int_C g(z)\, e^{\tau f(z)}\, dz,$$

where $g(z)$ is a slowly varying function (as compared to $e^{\tau f(z)}$. In this case, we replace $g(z)$ by its approximate value $g(z_0)$ in the integrand at a saddle point $z = z_0$. The remaining process of deforming the contour and evaluation of J remains unaltered.

8.4 Multivalued Functions and Riemann Surfaces

In Chapters 6 and 7, we have seen the development of the theory of analytic functions which includes several results for these starting with and following from the Cauchy–Gursat theorem. If we were to try to generalize these blindly to the multi-valued functions, we encounter an obstacle in this development. In the case of a single-valued function $f(z)$, a point z_0 in the complex plane is associated with a single complex number $f(z_0)$. This property of single-valued-ness is essential before one can define the derivative of a function, which first of all requires one to evaluate $[f(z_0 + \Delta z) - f(z_0)]$. This has a meaning only when each of terms has a unique value. Unless the notion of the derivative is defined, the concept of analyticity and the series of results based on it would not be possible. For example, the function $\log z$ has an infinite set of values at the point $\sqrt{2}(1+i)$ given by

$$\log\, z = \text{Log}\, 2 + i\pi/4 + 2\pi n i \qquad n = 0, \pm 1, \pm 2, \pm 3, \ldots\ldots$$

Similarly, the function $\sqrt{z}$ has two distinct values at, say, $\sqrt{2}(1+i)$:

$$\sqrt{z} = \pm\sqrt{2}\exp(i\pi/8)$$

while the function z^{π} has the following infinite number of values at $\sqrt{2}(1+i)$:

$$z^{\pi} = \exp\{\pi(\text{Log}\, 2 + i\pi/4 + 2n\pi i)\}.$$

We would like to make it possible to generalize the theory of analytic functions also to such multi-valued functions. We have already seen one way to do this: it was to choose a particular set of values, such that the function was single-valued over the complex plane and analytic over as large a domain as possible. For example, we had defined the *principal value* of the log function:

$$\text{Log}\, z = \text{Log}\, r + i\Theta \quad \text{with} \quad -\pi < \Theta \le \pi.$$

As seen in Chapter 6, the above function is single-valued and analytic except on the negative real axis (including origin) where it is discontinuous. Here, one has chosen a branch-cut along the negative real axis. In this way however, one is restricted to one particular branch at a time.

We bypass the above problem by constructing what we call a "Riemann surface" which may be thought of as a generalization of the notion of the complex plane. A Riemann surface is needed only for a multi-valued function. It may be distinct for each multi-valued function. (Moreover, there can be more than one construction of a Riemann surface for a given function, as we shall see). One construction of Riemann surface for $\log z$ is as follows: We imagine a surface, something like a circular staircase, obtained by taking infinite number of copies of the complex plane with a branch-cut along the negative real axis (called Riemann sheets). We then join them successively in the manner explained below.

Let us number the sheets by an integer 'n' which takes all integer values. For each particular sheet, we assign a specific range to the (multi-valued function) arg z. On the sheet number n, we have $(2n-1)\pi < \arg z \leq (2n+1)\pi$. Now, we join the infinite number of sheets successively such that the variable arg z varies continuously across the sheets even as one crosses the negative real axis. To be specific, we join the part of zeroth sheet below the negative real axis to the part of the $(-1)^{\text{st}}$ sheet above the negative real axis; the part of the zeroth sheet above the negative real axis to the part of the 1st sheet below the negative real axis, etc.

In this manner, as a point moves along a contour on the Riemann surface, (not passing through the origin) we find that the function $\log z$ is

(a) single valued at each point and
(b) analytic along the contour.

For example, compare what happens if we move anticlockwise along the path $r = a$ in

(i) the complex plane *and*
(ii) the Riemann surface constructed above.

In the first case, the path is a circle and the function $\log z$ is discontinuous as we cross the negative real axis. In the second case, the path is (analogous to) a helical one and $\log z$ is analytic along it.

The principal branch of $\log z$ lies, in fact, in the zeroth Riemann sheet. If we try to continue it analytically across the negative real axis, then

(a) continuing from the 2nd quadrant to the 3rd, we end up on the *first Riemann sheet*, while
(b) continuing from the 3rd quadrant into the 2nd in a clockwise sense, we end up on the *Riemann sheet* labelled by $n = -1$.

Thus, if we were to start from the principal branch $\log z$ and continue analytically across to branch cut both clockwise and anticlockwise in this manner indefinitely, we would generate the function $\log z$ on the entire Riemann surface.

It should however be remarked that for a given function (such as $\log z$ in the above example) there is no unique way of choosing the Riemann surface in as much as there is no unique way of choosing the branch cut. What is presented above is only one possibility. We could alternately choose the branch-cut for $\log z$ as:

$$\log z = \log r + i\theta \qquad \alpha < \theta \leq \alpha + 2\pi$$

where α is an arbitrary real number. We could then choose the nth branch as:

$$\log z = \log r + i\theta \quad \alpha + 2n\pi < \theta \leq \alpha + 2(n+1)\pi \qquad -\infty < n < \infty.$$

We could then construct a different Riemann surface by joining this infinite collection branches very much in the same manner. This will then be another choice of a Riemann surface for the same function.

Solved problem 8.1 Construct a Riemann surface for $f(z) = z^{1/5} \equiv r^{1/5}(\cos\theta/5 + i \sin\theta/5)$.

Solution: $f(z)$ is a multivalued function with 5 distinct values for any $z(\neq 0)$, obtained by replacing θ by $\theta + 2n\pi$; $n = 0, 1, 2, 3,$ and 4. We can choose five Riemann sheets as follows: On the nth *sheet*, the function is given by,

$$f(z) = r^{1/5}(\cos\theta/5 + i \sin\theta/5) \quad \text{with} \quad 2n\pi \leq \theta < 2(n+1)\pi;\ n = 0, 1, 2, 3, 4.$$

For a given $z \neq 0$, each sheet has one of the values of $f(z)$. The Riemann surface is then constructed by joining these in such a manner that as one moves on it along $r = a > 0$ in an anticlockwise sense; θ varies continuously from 0 to 10π and comes back to zero.

Problems

On §8.1

8.1 Consider the function defined over the domain $D_1 \equiv \{|z+1| < 1\}$

$$f(z) = -\sum_{n=0}^{\infty}(z+1)^n \qquad |z+1| < 1.$$

(a) Sum the series in D_1 and locate its singular points.

(b) One wants to continue the function analytically by using the form of $f(z)$ above and writing down Taylor expansion about the point $z_0 = -1/2 + 1/2i$. Find the largest circular domain D_2 (overlapping with D_1) to which $f(z)$ can be continued by this method.

(c) What should be the point z_1 within D_1 so that Taylor expansion about z_1 leads to an analytic continuation of $f(z)$ to a largest possible domain? Specify this domain. (Note: z_1 can only be specified in a limiting manner).

(d) Write down, with due arguments, the analytic continuation of $f(z)$ to the largest possible domain D and specify D.

8.2 Consider the function $f(z)$ defined over D_1 given below

$$f(z) = r^{1/2} e^{i\theta/2}; \quad r > 0, \quad \pi/2 < \theta < 3\pi/2$$

(a) Continue the function $f(z)$ analytically

(i) to D_2 given by $\{r > 0, -\pi/2 < \theta < \pi/2\}$, and
(ii) to D_3 given by $\{r > 0, 3\pi/2 < \theta < 5\pi/2\}$.

(b) Compare the analytic continuations so obtained to the two *overlapping* domains D_2 and D_3.
(c) Show that the idea of uniqueness of analytic continuations of $f(z)$ can be restored in this example if a suitable Riemann surface is introduced.

8.3 (a) Obtain the domain of definition of $I(z)$ defined below:

$$I(z) = \int_0^\infty e^{-(z+a)t}\, dt$$

(b) Using an argument that employs the uniqueness theorem, write down the analytic continuation of $I(z)$ to the entire complex plane (except for possible isolated singularities).

8.4 (a) Argue that $\sin z = (e^{iz} - e^{-iz})/2i$ is the unique entire function that coincides with $\sin x$ on the real axis. Now given that $\cos z$ and $\sin z$ are (the unique) entire functions that coincide with $\cos x$ and $\sin x$ over the real axis,

(b) show that

$$\cos(x_1 + x_2) = \cos x_1 \cos x_2 - \sin x_1 \sin x_2$$

implies

$$\cos(z + a) = \cos z \cos a - \sin z \sin a; \quad a \text{ real};$$

(c) using the above result show further that

$$\cos(z + Z) = \cos z \cos Z - \sin z \sin Z,$$

z and Z being arbitrary complex numbers.

(Hint: In (b), consider $f(z) = \cos(z + a) - \cos z \cos a + \sin z \sin a$. In (c), consider $f(Z) = \cos(z + Z) - \cos z \cos Z + \sin z \sin Z$; with z fixed.)

8.5 Find the domains of the definition of the following four integrals and show that they represent the same analytic function in the respective domains of definitions :

(a) $I_1(z) = \int_0^\infty dt\, e^{-zt}$

(b) $I_2(z) = i \int_0^\infty dt e^{izt}$

(c) $I_3(z) = z \int_0^\infty e^{-z^2 t}\, dt$

(d) $I_4(z) = -\int_0^\infty dt e^{zt}$

8.6 Let a real function $f(x)$ defined over $-a < x < a$ be infinitely differentiable in this interval.

(a) Show then that the function $F(z)$ defined through

$$F(z) \equiv \sum_{n=0}^{\infty} f^{(n)}(0)\, \frac{z^n}{n!} \qquad f^{(0)} \equiv f(0)$$

converges in the disc $|z| < a$ and represents a unique analytic function in this domain such that $F(x) = f(x)$; $-a < x < a$.

(b) Show that it satisfies the "Schwartz reflection principle"

$$F(z)^* = F^*(z)$$

within this domain.

8.7 Consider a function $f(x)$ of a real variable x which (i) is differentiable everywhere, (ii) equals x in [0, 1]. Show by explicit construction that this does not uniquely fix $f(x)$.
[Hint: You may for example construct two functions which differ (only) in $(1, \infty)$ but have same $f'(1)$.]. The contrast with a function of complex variable should be noted.

On § 8.2

8.8 Consider the domain D_1 to be the UHP in the complex z-plane. Consider the transformation

$$w = z^{1/3} \qquad -\pi < \arg w \leq \pi$$

(a) Find the domain D_2 in the complex w plane that D_1 is mapped into.

(b) Consider

$$f(z) = az; \qquad a \text{ real}$$

Consider $\phi(x, y) = \text{Re } f(z)$. What electrostatics problem is $\phi(x, y)$ a solution of?

(c) Consider the function

$$h(w) = az = aw^3 \equiv a(X + iY)^3$$

Show explicitly that the real and imaginary parts of $h(w)$ satisfy the Laplace's equation in the X–Y plane.

(d) Re $h(w)$ and Im $h(w)$ represent the potentials of certain electrostatics problems. Describe both of these.

8.9 Consider two curves C_1 and C_2 in the complex z-plane given respectively by $y = a$ and $x = a(a > 0)$. Consider the mapping $w = z^2$.

Find the mapping of C_1 and C_2 (viz. C_1' and C_2') in the complex w plane. Find the point of intersection of C_1' and C_2' and the angle between tangents to these at this point. Comment on the last result.

8.10 Two coaxial conducting infinite cylinders of radii r_0 and $1 (r_0 < 1)$ are given charges λ and $-\lambda$ per unit length.

(i) Show that the potential in the interior is given by

$$\Phi(\rho, \phi) = \frac{\lambda}{2\pi \epsilon_0} \ln \rho \qquad -\pi < \phi \leq \pi.$$

(ii) Determine the complex potential function $f(z)$.

(iii) Perform the analytic transformation

$$w = -i \text{ Log } z$$

(iv) Find the curves in the complex-w plane into which the two cylinders (i.e., their cross-sections) are mapped.

(v) Describe the new electrostatic problems represented by

$$\Phi_1(X, Y) = \text{Re } f(z[w])$$

$$\Phi_2(X, Y) = \text{Im } f(z[w])$$

8.11 Show that the integrand $e^{-t}t^w$ in § 8.3.1 is substantial (within 1% of the peak value) in the range given approximately, for large w, by

$$\left[w - \sqrt{2(\log 100)w},\ w + \sqrt{2(\ln 100)w}\right]$$

and thus that for $w = 1000$, the integrand is negligible outside the interval (0.904, 1.096).

8.12 From Problem 8.11 or otherwise, show that the integrand in Eq. (8.49) of § 8.3.1 is peaked sharply around $x = 1$ with width of the peak decreasing as $\frac{1}{\sqrt{w}}$ for large w. Show further that the Gaussian approximation Eq. (8.52) to this integral also implies the same.

8.13 $\Gamma(w+1)$ of § 8.3.1 Eq. (8.46) can be shown to satisfy $\Gamma(w+1) = w\Gamma(w)$. (see the accompanying volume *Mathematical Physics: Advanced Topics*). Assuming this relation and an expansin of the form

$$\Gamma(w+1) = e^{-w} w^{w+1/2} \sqrt{2\pi} \left\{ 1 + \frac{A}{w} + 0\left(\frac{1}{w^2}\right) + \cdots \right\}$$

show that $A = 1/12$; thus the leading result is accurate within 1% for $w \geq 8$ when compared with the next leading order. [Hint: The binomial expansion for $(w-1)^w$ can be obtained as that for $w^w e^{w \log(1-1/w)}$].

8.14 Obtain the leading behaviour of

$$I(w) = \int_0^\infty dx e^{iw(x-x^2/2)}$$

for large w by the steepest descent method. Justify assumptions made.

8.15 Modified Bessel functions (related to Bessel function J_ν with imaginary arguments) $I_\nu(x)$ are given in terms of a contour representation

$$I_\nu(\zeta) = \frac{1}{2\pi i} \int_C e^{\frac{\zeta}{2}(z+\frac{1}{z})} z^{\nu-1} dz$$

where the contour starts and ends at $z \to -\infty$ circling the origin in the positive sense. Determine the asymptotic behaviour of $I_\nu(\eta)$ as $\eta \to \infty$.

Noting $J_\nu(\eta) = i^\nu I_\nu(-i\eta)$, evaluate the asymptotic form of $J_\nu(\eta)$ as $\eta \to \infty$. Compare the deformation of contours through saddle points.

8.16 Complete the evaluation in the example of Section. 8.3.2.

Index